The Renaissance of Renewable Energy

GIAN ANDREA PAGNONI

STEPHEN ROCHE

UNIVERSITY PRESS

32 Avenue of the Americas, New York, NY 10013-2473, USA

Cambridge University Press is part of the University of Cambridge.

It furthers the University's mission by disseminating knowledge in the pursuit of education, learning and research at the highest international levels of excellence.

www.cambridge.org
Information on this title: www.cambridge.org/9781107698369

© Gian Andrea Pagnoni and Stephen Roche 2015
Cartoons provided courtesy of Roberto Malfatti (http://rmalfatti.blogspot.it)

This publication is in copyright. Subject to statutory exception and to the provisions of relevant collective licensing agreements, no reproduction of any part may take place without the written permission of Cambridge University Press.

First published 2015

Printed in the United States of America

A catalog record for this publication is available from the British Library.

Library of Congress Cataloging in Publication data
Pagnoni, Gian Andrea.
The renaissance of renewable energy / Gian Andrea Pagnoni, Stephen Roche.
  pages cm
Includes bibliographical references and index.
ISBN 978-1-107-02560-8 (hardback) – ISBN 978-1-107-69836-9 (paperback)
1. Renewable energy sources. 2. Energy consumption–Environmental aspects.
I. Roche, Stephen. II. Title.
TJ808.P33 2014
333.79′4–dc23   2014026000

ISBN 978-1-107-02560-8 Hardback
ISBN 978-1-107-69836-9 Paperback

Cambridge University Press has no responsibility for the persistence or accuracy of URLs for external or third-party Internet Web sites referred to in this publication and does not guarantee that any content on such Web sites is, or will remain, accurate or appropriate.

Everything in the universe may be described in terms of energy. Galaxies, stars, molecules, and atoms may be regarded as organizations of energy. Living organisms may be looked upon as engines which operate by means of energy derived directly or indirectly from the sun. The civilizations, or cultures of mankind, also, may be regarded as a form or organization of energy.
Leslie White, 1943

Contents

	Preface	page ix
	Introduction	1
1.	What Is Energy?	3
2.	Where Does Energy Come From?	26
3.	How Much Energy Is Enough?	45
4.	How Energy Is Produced	69
5.	Challenging Times: The Politics and Economics of Energy	171
6.	The Price of Energy Consumption	189
7.	Energy from My Backyard	232
	References	273
	Index	287

Preface

This book began with a skiing trip in the Alps in January 2010. Gian Andrea and I had known each other since 1990, when, as students, we shared an apartment in Ireland. Having lost touch for more than a decade, we had recently renewed the friendship. Conversations between old friends tend to cover a lot of ground very quickly. We filled each other in on relationships past and present, on travels and travails, past and current work, and where we thought we were heading. I'd been working as a translator and editor for several years. Gian Andrea had set up an environmental consulting company with other scientists from the University of Ferrara and balanced this work with additional teaching and writing.

He had always been a practical man. In kitchen table conversations during our student days, he didn't shy away from controversial positions, exposing the contradictions within many closely held beliefs. He once declared, to the horror of several humanities' students, that he felt no attachment to language, and that the disappearance of his native Italian would be of no consequence provided it were replaced by a more efficient substitute, such as, for example, English. I was bemused and a little shocked by his lack of sentimentality, but I appreciated his dedication to objectivity.

At the time of the skiing trip I was starting to tire of my ventriloquial craft, so when he suggested an alternative, I was curious. "If you're tired of working on other people's books, you should write your own. In fact, we could write one together." I had recently attended the UN Climate Change Conference in Copenhagen, so the point where my interests and his expertise most clearly intersected was obvious: sustainable energy.

The process began with a series of Skype conversations, leading to a proposal we jointly composed using Google Drive. As the project progressed, Skype and Google became indispensable tools, allowing for a kind of collaboration that would not have been possible without them. Though we were living about 1,000 kilometres apart, most evenings, for a couple of hours, we stepped into the same virtual room. Most of the sections began with a conversation, during which Gian Andrea introduced a particular idea or data, and I, by a kind of Socratic process, teased out what that meant in layman's terms. Once we felt we had arrived at an explanation that conveyed the essence without compromising the scientific fact, we composed the particular section. Eventually, we corrected these together and discussed figures, captions and changes. We probably wrote ten times the number of words currently in the book. The result is neither the work of a scientist simplifying things for the layman nor of the layman coming to grips with science; rather, it is a marriage of the two.

Stephen Roche

Introduction

Renewable energy, far from being a new idea, was the norm for most of human history. The ability, developed in the late eighteenth century, to harness the 'fossilised' energy of coal and oil on a large scale transformed human societies. It was no accident that the explosion of the Earth's human population, from roughly 1 billion in 1800 to more than 6 billion in 2000, coincided with this energy revolution. The question people have been asking in increasing numbers for the last fifty years is, "Can it go on like this?" Most are in agreement that it can't. Whether one views climate change, population growth or resource depletion as the greatest threat to human survival, the basic problem is the same: there are limits to what our planet can provide or absorb. The renaissance of renewables is inevitable because sooner or later the oil, gas and coal will run out; because by releasing in decades the carbon absorbed over millennia we are choking the planet; and, lastly, because of economics – whereas fossil fuels are likely to become more expensive over time, renewables can only get cheaper.

This book covers most of the issues related to renewable and sustainable energy – from the purely technical to the historical, political, social and economic. It begins with a broad introduction to energy, both as a concept and a practicality, outlining the history of human energy use, profiling our current consumption, and assessing likely prospects for the future. In doing so, it uncovers and unpacks the various factors that influence our energy choices.

This book is also about the transition we must make, and indeed have already begun to make. It explains the different factors influencing that transition and the likely sacrifices that will be required. Stephen lives in Germany, so we are familiar with the debates that led to the adoption of the 'Energiewende' (literally, energy transition) in 2011 in the aftermath of the Fukushima nuclear accident in Japan.

This created a critical mass of support among the German public for a departure from nuclear energy. At that time Germany had only seventeen nuclear power plants in operation, supplying about 18 per cent of the nation's electricity. As these are phased out, they are mostly being replaced in the short term by brown coal (of which Germany has plenty) and Russian natural gas, which at the time of writing entails a precarious dependency. Renewables supply about 23 per cent of Germany's electricity and about 14 per cent of its overall energy consumption.[1] These figures may impress when compared with those of most other countries, but on their own they show that the 'energy transition' is still in its infancy.

The transition to sustainable energy will not only transform the way energy is generated but also the way it is traded. Already, the model of electricity supply devised by Thomas Edison in the early twentieth century – a relatively small number of very large power stations that supply power via a ubiquitous grid – is starting to appear outdated. Whereas a generation ago, many energy utilities prospered by operating large power plants of one particular type, many of these giant companies are now investing in a wide range of alternatives. In the United States alone, there are now close to 500,000 solar plants in operation. Most of these are small and are installed on the roofs of private homes. A quarter of them were installed in 2013 alone (Biello 2014).

Every human intervention in the natural environment has an impact. In the cases of hydropower and bioenergy, the impacts may even exceed – in terms of the environmental and social disruption – those of fossil or nuclear power. This book therefore is not biased in favour of renewables but considers the price of the 'energy transition' in terms of environmental and social impacts as well as economics.

Every movement begins with an idea, and once an idea has been widely embraced, change can follow quickly. We believe that the most immediate obstacle to a peaceful energy transition is not economic, infrastructural or political. It lies in the ability of large numbers of people to not merely reject the existing system but to imagine a new paradigm. We hope that this book can help to fill some of the gaps in your understanding of energy, and help you develop a clearer idea of how the energy transition can occur.

Gian Andrea Pagnoni

---

[1] Source for all German energy statistics: German Association of Energy and Water Industries (Bundesverband der Energie- und Wasserwirtschaft [BDEW]) http://www.bdew.de/.

1
What Is Energy?

1.1 Aristotle in Times Square

The term 'energy' has become ubiquitous, as likely to be heard in a yoga class as at a physics lecture. In its everyday use, it has become synonymous with force, vigour, well-being and a certain kind of atmosphere. We talk about people or places having energy, a certain kind of energy, or lacking it altogether. We've become so used to using the words 'energy' and 'energetic' as pliant descriptors that we're liable to overlook their scientific significance.

A first-time visitor to Times Square, the heart of one of the world's busiest cities, is likely to first comment on the 'energy' of the place. But does this use of the term bear any relation to its scientific meaning? The Greek term ἐνέργεια (*energeia*), the origin of the English word, was probably coined by Aristotle. It combines the prefix *en*, meaning 'in' or 'at', with *ergon*, meaning 'action' or 'work'. According to Aristotle, all living beings are defined by this attribute; they are 'at work', in contrast to inactive, inanimate objects. So *energeia*, for Aristotle, was intimately connected to movement. This philosophical concept of energy remained for more than 2,000 years the main usage of the term. As late as 1737, the philosopher David Hume wrote that there were "no ideas, which occur in metaphysics, more obscure and uncertain, than those of power, force, energy or necessary connexion".

The first attempts to define energy in scientific terms date back to the seventeenth century. Isaac Newton established that the same force (gravity) which causes an apple to fall from a tree also determines the movement of the planets around the sun. Newton's contemporary Gottfried Leibniz identified something he called *vis viva* (literally 'living force'), the force of any moving thing. Leibniz began the process of quantifying energy when he concluded that while the force of a moving object depends both on its mass (weight) and its velocity (speed), velocity was far more important than mass. In other words, a light but fast-moving object has far more force than a heavy but slow-moving one. Just imagine catching a basketball, which weighs about 600 grams, thrown by a teammate. Now compare this with the impact of a 10-gram bullet fired from a gun.

The human understanding of energy took a huge leap forward during the Industrial Revolution, pioneered by industrialists who were motivated as much by commercial ambition as by scientific enquiry. For them, energy was not an abstract idea; it was the force needed to drive the machines that were rapidly replacing human and animal labour. They therefore redefined energy as the ability to perform work. This remains the most common definition to this day. But what exactly do we mean by work? An ox pulling a plough is clearly at work. The animal's 'biological' energy is converted into furrows. In scientific terms, the ox exerts a force over a distance. Since prehistoric times, humankind's work, like that of the ox, has mainly involved moving objects, whether spears, arrows, goods or the plough. By the mid-eighteenth century, it was the turn of machines, and in order to build and use those machines, people needed to understand and quantify energy.

Most work requires more than the mere application of energy. To be effective, that energy must be concentrated. We see this when we open a bottle of beer or a soft drink. It would take a very strong (and thick-skinned) person to tear the cap from the bottle without using a tool. However, even a young child can perform the same task with a bottle opener. This is because the opener works as a lever, concentrating the energy at the point where it is needed to remove the cap. When energy is concentrated not in terms of space (such as at the rim of a bottle) but in terms of time, the concept of power comes into play. Most people have gone through the ordeal of moving house at least once. If we do the move ourselves, the time required will depend largely on the muscle power we can muster from obliging friends and family members. If you have a few bodybuilders in the family, the move will be quick. If you are relying mainly on your kids, you should hire the van for the entire week. This, essentially, is the difference between energy and power: power is the *rate* at which energy is generated and consumed to perform work.

James Watt (1736–1819) was particularly interested in power. He spent most of his life improving the steam engine, which works by heating water to form steam. The vapour occupies a greater volume than liquid water does and so pushes upwards, raising a piston, just as water boiling in a pot raises the lid. Thus, Watt (and others before him) succeeded in converting the energy of heat into the energy of movement, which can be harnessed to perform a wide variety of tasks, from pumping water to turning a wheel. To convince his customers of his machine's efficacy, Watt came up with the term 'horsepower', which explains its power output relative to the main energy source of his day, the draft horse. This term, which is still used to rate certain types of engines, was later, fittingly, replaced by the *watt* as the international unit of power.

Converting Energy

Watt's horsepower measured the output of his machines, but it fell to another entrepreneur-engineer to measure the transformation of one form of energy to another. While exploring ways to improve his brewery, James Joule (1818–1889) made a breakthrough. He had been thinking of changing over from the steam engine to the newly developed electric motor. Before doing so, he wanted to compare the amount of work that could be performed by each machine. Joule constructed a device resembling an egg beater immersed in a jar of water, and he used a weight and pulley to turn the blades of the 'beater' (see Figure 1.1). The movement of the water molecules created heat, which Joule was able to measure using a thermometer. The greater the weight (force) he used, the faster the beater turned, and the greater the rise in temperature. In this way, he discovered a simple yet remarkably accurate way of measuring the relationship between work and heat.

Joule's experiment led to the formulation of one of the most important principles of physics: the first law of thermodynamics. This states that energy can be neither created nor destroyed, but merely changed from one form to another. Think of what happens when a car brakes: the energy of its movement is not lost, just converted into another form of energy. The brake pads, discs and surrounding air are warmer than they were before the driver braked. This principle is crucial to understanding how energy can be generated and used.

The Enigma of Energy

By the twentieth century, scientists had learned to quantify and measure energy, yet there remained something inherently mysterious

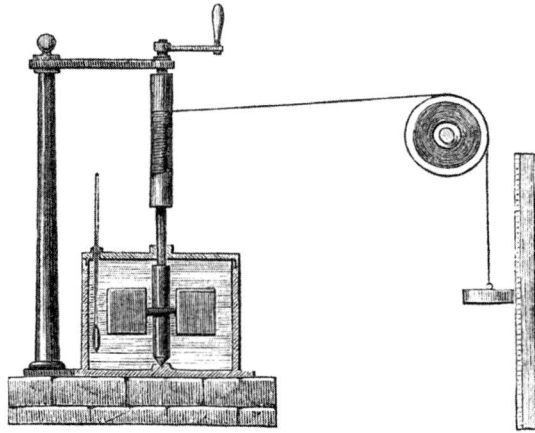

Figure 1.1. Joule's apparatus for measuring the relationship between work and heat. The fall of the weight causes the blades to turn, stirring – and thus heating – the water inside the container (calorimeter). A thermometer measures the rising temperature.

about the concept. Richard Feynman, one of the towering figures of modern physics, went as far as to admit that "in physics today, we have no knowledge of what energy is" (1970).

Energy, as we currently understand it, is force, work and power. It is at the heart of what it means to be alive: the ability to manipulate our environment to meet our needs. Thus, Aristotle's definition of energy remains essentially valid today. If the father of Western philosophy were to have stepped into a time machine that touched down on Times Square, he would recognise around him the principle of energy in action, through two factors: motion and work.

1.2 Energy: What Gets Lost in 'Translation'

Robert Frost memorably defined poetry as "what gets lost in translation." Just as meaning is inevitably lost as ideas are converted from one language to another, there is no way to convert energy without loss. Energy efficiency, like translation, is merely about minimizing that loss.

As a teacher, I always require that my students work in groups, where each person's grade is influenced by that of the whole group. It often happens that a student complains about a group-mate, typically that he or she is not pulling their weight and therefore jeopardising the performance of the group. This reflects the second principle of thermodynamics. The second principle states that an enclosed system naturally tends towards maximum disorder, or entropy. Mountains are gradually worn down by wind and water, houses need to be regularly repaired and maintained, and teenage students have a gift for creating mayhem. Being a good student means applying a great deal of order to one's behaviour. Yet this is an energy-consuming process. Sometimes my students opt for a strategy that involves a much lower energy investment; instead of studying and supporting each other, they try to sow doubt in my mind and blame each other, thus creating disorder in the classroom.

Students are not the only ones affected by entropy. All living beings expend an enormous amount of energy every day of their lives just maintaining the status quo. Order is needed to survive and triumph, at least for a while, over the many external forces that are out to get us. To live we must actively counteract the second principle, and this requires that we expend energy. The second principle implies not only that it is far easier to destroy (creating disorder) than to build (creating order) but also that any conversion will inevitably entail some dissipation of energy, usually in the form of heat. Strictly speaking, the energy converted into heat has not been lost. However, it is not easily recovered. Staring into an open wood fire on a cold winter's evening, it is easy to become mesmerised by the sparks rising with the smoke. What we are witnessing is the chemical energy stored in the wood being converted into heat and light. However, the second principle prevents the opposite occurring: heat and smoke cannot be converted back into a woodpile. Part of the energy has been so widely dispersed that it cannot be retrieved.

The Low Efficiency of Energetic Conversions

Strictly speaking, the terms 'energy production' and 'energy loss' are incorrect, as – according to the first law of thermodynamics – energy can be neither created nor destroyed. What we observe in physics or chemistry is merely a conversion from one form of energy into another. Fully efficient energy conversion is possible only in theory, and indeed most conversions are highly inefficient. The engine of a car provides a good example of how energy gets 'lost' in conversion. Cars run thanks to a controlled explosion in the combustion chamber. Thus, chemical energy (fuel) is first converted into thermal energy (heat), and then into kinetic energy (motion). However, within

Table 1.1. *Comparison of different forms of energy conversion and their efficiencies*

Process/technology	Conversion	Efficiency	Energy loss
Photosynthesis (wild plants)	light → chemical bonds	0.2–0.3%	heat
Photosynthesis (crops)	light → chemical bonds	2–5%	heat
Muscles	chemical → movement	30%	heat
Candle	chemical → light	0.01%	heat
Candle	chemical → heat	99.99%	light
Incandescent light bulb	electric → light	10%	heat
LED lamp	electric → light	50%	heat
Steam engine	chemical → heat → movement	5%	heat, noise
Electric engine	electric → movement	80%	heat, noise
Car (internal combustion engine)	chemical → heat → movement	10%	heat, noise
Gas turbine	chemical → heat	> 95%	noise
Gas turbine	chemical → heat → mechanical	60%	heat, noise

this threefold conversion process only 10 per cent of the chemical energy contained in the petrol or diesel is converted into motion. So, what happens to the other 90 per cent? About three-quarters of it is lost either as heat or consumed by the car's cooling system, while the remainder is lost as a result of friction (of tyres gears and air drag), idling, and auxiliary functions such as air-conditioning and power steering. Some conversions are even more inefficient (for example, a candle transforms no more than 0.01 per cent of the chemical energy in the wax into light), while others are considerably more efficient: an electric motor transforms about 80 per cent of the electricity consumed into mechanical energy.

1.3 The Various Forms of Energy

Consider for a moment what it takes to read these lines. First, there is the energy required to maintain a constant body temperature, then that used by the movement of the eye, and finally the energy required by the brain to process the message. At no moment in our lives do we cease to expend energy. Even during sleep the human body performs a variety of tasks: the heart beats; blood circulates; enzymes and hormones digest, protect, repair, and maintain temperature; and the brain, our most energy-intensive organ, works to maintain control of the body. Like the human body, the world partakes in a constant exchange of energy. It is, in the words of energy expert Vaclav Smil, "the only universal currency. One of its many forms must be transformed into another in order for stars to shine, planets to rotate, living things to grow, and civilizations to evolve" (Smil 2000).

Gravity is perhaps the first type of energy we experience in life. As a baby emerges from her mother's womb, she experiences for the first time a sense of weightedness. This force – no doubt disconcerting to a newborn – is truly universal. All bodies in the universe, from atoms to stars, exert a gravitational attraction on one another. This force is directly proportional to the mass of the attracting body and indirectly proportional to the distance from it. That is why astronauts can bounce around on the moon like slow-motion trampolinists (our satellite has one-fourth the mass of our planet) and why we are drawn to the Earth rather than the sun (the sun's mass is more than 300,000 times that of the Earth, but it is 150 million kilometres away).

A falling object, attracted by gravity, exerts another form of energy: kinetic energy. This can be transferred from one moving object to another, as when a tennis racket strikes a ball. However, not all the energy is converted in this way. Because the atoms within the tennis ball are excited and vibrate, they generate heat, or thermal energy. Heat is therefore a form of kinetic energy, generated at the atomic level.

Heat can be transferred either by the physical impact of particles or in the form of electromagnetic waves. We are familiar with mechanical waves; by their nature they are tangible – whether as sound travelling through air, waves in the ocean or ripples in a pond. Yet electromagnetic waves are an equally constant and natural feature of our world, in the form of radio waves, microwaves, X-rays, and gamma rays. Light is one such electromagnetic wave. Heat also radiates, in the form of infrared rays that can be "seen" by some species of snake through special thermal receptors.

The ancient Greeks found that amber, when rubbed against animal fur, exerted an attraction on small objects. As a result of this discovery, the Greek term for amber (*elektron*) forms the root of the

English word 'electricity.' Electricity describes the presence and flow of electrons, tiny negatively charged particles that orbit the nucleus of every atom. Manifestations of electricity include lightning, static electricity and the flow of electrical current in a copper wire. Certain elements, particularly metals, easily release and receive electrons. When we flick a light switch or turn on an appliance, we take advantage of a flow of electrons jumping from atom to atom along a copper wire, a flow that began, in most cases, at the nearest power plant.

A chemical reaction occurs when one chemical element 'donates' electrons to another. The fascination and comfort many of us feel while staring into a campfire may be attributable to the fact that combustion is humankind's oldest source of external energy. A typical combustion reaction sees carbon react with oxygen, releasing carbon dioxide (CO_2), water (H_2O), and energy in the form of heat and light.

Every chemical transformation is accompanied by an increase or decrease in energy. In order to lift a book from the floor onto a table, we need to expend energy; the muscles of our arms convert some of the chemical energy we consumed as food into mechanical energy. To raise the book even higher onto a bookshelf, we must expend even more energy. The floor, the table and the bookshelf represent three energetic levels. If the book falls from the shelf, the energy we invested in it will be released in kinetic and thermal energy, as the molecules in the air and the floor are excited. Because of this, we say that the book on the bookshelf has potential energy.

There are numerous ways to store energy. For example, electric energy may be stored in a battery and kinetic energy behind a dam. The electrons in the battery and the water molecules behind the dam are 'poised' to release energy. The sum of potential and kinetic energy is known as mechanical energy. This is the energy associated with the motion or position of an object. The classic example is a swinging pendulum. The pendulum passes back and forth between kinetic and potential energy. It attains its maximum kinetic energy and zero potential energy in the vertical position, because it reaches its greatest speed and is nearest the Earth at this point. At the extreme positions of its swing, on the other hand, it will have its least kinetic and greatest potential energy. The energy never leaves the system but is constantly converted between kinetic and potential. The pendulum slows down and eventually stops only because a part of the energy is converted into heat through air drag and friction at the pivot.

1.4 Qualities of Energy

Earthquakes and lightning are among nature's most dramatic shows of force. Little wonder that so many cultures have constructed myths and legends around these phenomena, imagining wrathful gods venting displeasure and exacting vengeance. Even without a divine interpretation, these are awe-inspiring events. The 2011 earthquake that in few minutes triggered floods and nuclear meltdown in Japan released enough surface energy to power the city of Los Angeles for a year. A typical lightning bolt releases about one million megawatts, enough to meet the electricity needs of Germany and France, though only for a fraction of a second! Yet we are unlikely to tap these immense sources of natural energy anytime soon. The energy of a lightning bolt is far too concentrated (any conductor or battery we can currently envisage would be fried to a crisp), and that of an earthquake far too dispersed to harness. Solar radiation, by stark contrast, strikes the Earth's surface with only 13 watts of power per square meter, yet is far more useful, as we can convert it relatively easily into electricity.

For energy to be useful, it must have at least one of three basic characteristics: it must be sufficiently but not excessively concentrated; it must be storable; and it must be transportable. Not all forms of energy (thermal, chemical, electrical, etc.) have the same level of usability, and only chemical fuels meet all three requirements. Electricity meets the first and the third conditions. Although it may be stored in sufficient quantity to power small appliances, such as phones, laptops and flashlights, electricity storage at an industrial scale is still a long way off. That is why hospitals rely on diesel generators rather than batteries in case of blackouts. Thermal energy may satisfy the first condition but can be stored and transported only at low intensity (e.g., in a thermos flask).

> **The Three Characteristics of Useful Energy**
>
> **Concentrated:** The tide is able to raise and lower every vessel in a port, but is of little help in raising even a small boat within a few minutes or maintaining that position for days. It is often possible to alter the concentration of an energy supply, but it is far more difficult to concentrate dispersed energy than to disperse concentrated energy.

(continued)

> **Storable:** A conventional diesel-engine car can run for about 700 kilometres on a single tank of fuel. This is the great advantage of fuel-based energy sources. Even the most efficient electric cars can only manage about a third of that distance on a single charge, and they require several hours to recharge.
>
> **Transportable:** Natural gas or electricity can be channelled through pipes and cables for thousands of kilometres, while losing very little of their energy. Heat, by contrast, is rapidly dispersed and requires thick insulation for transport even over short distances.

1.5 The Biology of Energy

When we humans compare ourselves to other species, it is usually to emphasise our more attractive qualities. Thus, a burly man may be compared to a lion or an oak, a beautiful woman to a swan or a rose. Few humans would feel flattered at being compared to a mushroom. Yet, in one very crucial respect, we have more in common with fungi than with an oak or a rose. We humans – like all other animals, all forms of fungi, and some types of bacteria – are heterotrophs. This means that we need to consume other organisms to survive. The oak and the rose, however – like all plants, all algae and some bacteria – are autotrophs. This means they are able to synthesize their own food (from the Greek *autós*, meaning self, and *trophē*, meaning nourishment). They use solar energy to power a reaction that converts water and carbon dioxide into organic biomass (sugar). We call this chemical process photosynthesis (from the Greek *photos*, meaning light).

The synthesis of highly ordered sugar molecules from unordered and scattered molecules of carbon dioxide and water could not happen spontaneously. That would contravene the second principle of thermodynamics, according to which all things tend towards decay and disorder. It is the input of solar energy that makes the transformation possible.

Once an organism has obtained its food, whether through photosynthesis or by consuming other organisms, it converts that fuel into energy. This process – essentially a kind of cellular combustion – is known as respiration. Oxygen absorbed by the lungs is distributed through red blood cells to every cell in the human body. In the cells, the oxygen reacts with the organic fuel to produce energy. The by-products of this reaction – water and carbon dioxide – are exhaled through the lungs and excreted through the kidneys. Plants respire

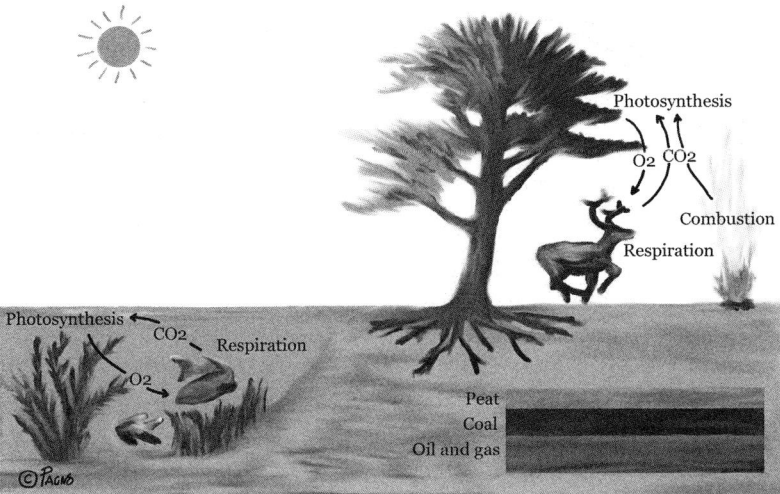

Figure 1.2. All life depends on the ability of some organisms to transform inorganic matter into organic matter using solar energy. The sun, therefore, is the first link in the chain of life.

in a similar way, releasing carbon dioxide and water through their leaves.

Many living organisms, including most animals, have a second way of converting food to energy: fermentation. Unlike respiration, fermentation does not require oxygen. In evolutionary terms, it is also far older than respiration, but around eighteen times less efficient. Not surprisingly, then, all multicellular organisms, from plants to fungi and animals, rely mainly on respiration. Organisms that depend entirely on fermentation include yeasts and some bacteria. The pungent odours that emanate from rotting food are caused by the by-products of fermentation (carbon dioxide, alcohol, lactic acid and acetic acid). All methods of food storage and conservation, from smokehouses to fridges, aim to prevent or slow down fermentation. Animals use fermentation as a kind of backup, when more energy is needed than can be supplied using respiration, usually because of an inability to breathe quickly enough. Humans also take advantage of fermentation to make numerous food products – including alcoholic drinks, yogurt, bread and vinegar.

Ultimately, all life on Earth is supported by the relationship between the sun and autotrophic organisms, as the energy stored in their cells is transferred to other organisms in the food chain. At the end of its life, every organism enters what is called the detritus chain: its biomass is decomposed by insects and bacteria. As a result, the complex organic molecules, full of chemical energy, that make up the organism return to the soil as simple inorganic molecules that can be recycled into biomass by autotrophs. In this way, all life is part of an energetic flux. The irony of our recent 'discovery' of recycling as an approach to waste management is that all the processes on which our lives depend involve the principle of recycling. When herbivores eat grass or carnivores eat herbivores, they consume part of the solar energy stored in biomass via photosynthesis. The atoms that comprise our body belonged just a few days ago to another living organism, and perhaps a few years ago to another human being.

1.6 From Rubbing Sticks to Splitting Atoms: The History of Human Energy Use

According to the American biophysicist Alfred Lotka, any organism or organic system will tend to increase in size and complexity as long as there is enough available energy. Mould, for example, will spread over a piece of moist bread as long as there is space available and organic material from which to recover energy. Once this energy source is exhausted, the mould dies and is itself consumed by bacteria. Humans, like all animals, are heterotrophs, obtaining our energy by ingesting organic compounds. The astonishing reproductive success of our species is largely attributable to our consistent ability to discover new sources of energy and to optimise their use.

Between seven and six million years ago the dominant vegetation in East Africa changed from woody forest to savannah. Formerly tree-dwelling primates were forced to adapt to life on open grasslands. Evolutionary scientists have identified this as the moment when the earliest stage of the human line (known as hominids) diverged from the evolutionary line of apes (Pollard 2009). The change in environment forced the first major evolutionary change in the history of humankind: the development of upright gait (Lovejoy 1988; McHenry 2009; Wong 2006). The major advantage of walking upright was that it freed up the front limbs for other uses. The evolution of the human hand with its opposable thumbs allowed us to grasp and hold objects, leading to another great evolutionary achievement: the manufacture and use of tools.

Scientists have interpreted the use of tools as a key indicator of intelligence, yet tool usage was probably as much a cause as an effect of greater brain power. As early humans experimented with more complex and intricate tools, they came to rely more on cognitive faculties and less on instinct and emotion. Over millennia, this resulted in evolutionary growth in the corresponding parts of the brain. However, a larger brain also entails a greater energetic cost. The human brain consumes about 400 kilocalories per day, three times that of a chimpanzee. The larger, more energy-intensive brain paid its way by allowing us to increase our energy budget, first and foremost through the ability to develop and use tools, which in turn helped us to obtain more food, build shelter and make clothing. Our more complex brains were therefore a success in evolutionary terms because they contributed to a net energy gain (Isler and van Schaik 2009).

While early humans displayed major differences of psychology and social organisation, from a purely energetic point of view they lived much like other mammals: to obtain food they relied exclusively on muscle strength and the elementary stratagems of hunting and gathering. The energy requirements of the hunt provided a powerful incentive to socialize. Since a lone hunter's chances of killing large animals were low, groups of hunters cooperated to pursue, trap, kill and slaughter an animal, and to transport the meat back to the settlement. As this was generally beyond the ability of a single family, hunter-gatherers formed larger social groups and shared the fruits of collective effort.

Those prehistoric societies that evolved amidst a plentiful food supply naturally saw a gradual increase in social complexity, up to the levels associated with the most advanced agrarian societies: permanent settlements, high population density, large-scale food storage, social stratification, elaborate rituals and early forms of cultivation.

Fire: The First External Energy Source

Tools allowed hominids to make the most of their muscle power by concentrating energy, but it was only when they learned to use and control fire that they were able to harness and manipulate an external source of energy. No other animal has done this. In this sense, the discovery of fire was a watershed in the evolution of humanity, marking the birth of what we might call *Homo energeticus* (Niele 2005). With fire, humans were able to cook food, warm their surroundings and keep dangerous predators and insects at bay. Firelight also extended the productive day in winter. It is likely that hominids discovered fire by accident, as a result of lightning or wildfires, and for several hundred millennia were able to use but unable to start it. This meant that fires had to be tended and maintained over generations.

From fire control to fire making.

The fossil record suggests that food may have been cooked as early as 1.9 million years ago, while the earliest reliable evidence for controlled use of fire dates to about 400,000 years ago. Roughly 100,000 years ago (David et al. 2009) humans learned to start fire at will, using either stones such as flint to create sparks or the friction of dry wood to create embers. By the start of the late Stone Age (about 40,000 years ago), human mastery of fire had advanced to the point of using lamps that burned animal fats (Smil 2006), and just a few millennia later humans were firing clay into pottery figures.

E + T = C

Twenty thousand years ago, the Earth was still in the grip of its most recent ice age. The polar ice caps extended southward to the latitudes of modern-day London and New York, and mammoth roamed the subarctic tundra of central Europe and Asia. Crouching in the sparse, wind-gnarled bushes, two groups of hunters, very different in gait and appearance, stared out into the open plains.

Humans and wolves competed for the same prey and both had social systems that enabled them to hunt in packs. They learned to fear and eventually respect each other, and they finally discovered the advantages of teaming up. Initially, this partnership was probably based purely on mutual advantage. For the wolf, the human use of weapons meant a share in a greater number of kills, and perhaps even an occasional taste of larger prey, such as mammoth. For humans, the wolf's speed and ferocity was the equivalent of a new weapon.

Not long after we began to keep and breed dogs as hunting partners, humans domesticated sheep and goats (ca. 9000 BCE), giving us a reliable source of energy in the form of meat and milk, and facilitating the move away from hunting and gathering to agriculture. Agriculture emerged independently in the Fertile Crescent, South Asia, Oceania, Africa's Sahel and several parts of the Americas, starting with the eight so-called Neolithic founder crops: emmer wheat, einkorn wheat, barley, peas, lentils, bitter vetch, chickpeas and flax (Brown et al. 2008). By 4000 BCE, agriculture was widely practised in many of the fertile regions of the world, and cattle, pigs, horse and dromedary camels had also been domesticated. Highly organised net fishing of rivers, lakes and ocean shores also brought in great volumes of food. So profound were the changes to human lifestyles brought by this new relationship to food energy, that anthropologists refer to this as 'The Neolithic Revolution'. Agriculture and the domestication of animals changed humans' energetic pathways and our cultural evolution. Not only did agriculture give us a stable and predictable food supply but, thanks to selective cultivation, it gave us varieties of plants with far higher energy yields than are found in the wild. A wild grain such as einkorn wheat converts only 0.3 per cent of the sunlight energy that strikes its leaves into biomass, while modern strains have conversion rates ten times higher.

Agriculture allowed humans to maintain a relatively stable environment, thus ensuring a more predictable future. It was this predictability that allowed us to develop complex social structures and culture. The reduced daily pressure to secure food allowed humans to devote time to other pursuits: the refinement of tools and language, the construction of more permanent settlements, and the development of complex social relationships. In this way, agriculture is the most essential of all cultural achievements. Were we to try and express this in an equation to rival Einstein's, we might say that E + T = C, or energy (through an abundant food source) plus time (through the predictability of that energy source) begets culture.

Playing with Fire

For millennia, wood, dung and crop residues were the main fuel sources for heating and cooking. Indeed, these are still important domestic fuels in many countries. Fire was, of course, used not only for cooking and heating but also as a source of light. Oil lamps (burning animal and vegetable fats) have been used since the Paleolithic age (up to 40,000 years ago), while the more practical and versatile candle, using plant and animal waxes, was developed about 2,000 years ago. These remained the principal method of lighting right until the early nineteenth century (Smil 2006).

One of the main reasons wood was eventually replaced as a major fuel was its value for other applications, particularly as a building material and as the raw material for charcoal. Wood is made of lignin, which consists of carbon, hydrogen and oxygen. By superheating the wood, we can remove the hydrogen and oxygen atoms in the form of water (H_2O), leaving pure carbon. This is a vastly accelerated replication of what happens in the earth when coal is formed, hence the name of this by-product: charcoal. The ability to make charcoal was developed about 5,000 years ago. The importance of this discovery is that it provided a fuel with a greater energy density than wood. Charcoal also produces far less smoke, so it is well suited for indoor cooking. But by far its most significant application was in the smelting of metals, which opened the way for vastly more effective tools and weapons.

From Human Power to Horsepower

The desire for a more stable source of food led humans to domesticate wild animals; sheep and goats initially, followed by pigs, oxen and chickens. Humankind also realised that it could take advantage of certain animals' superior strength to pull sledges, ploughs and wheeled wagons. Thus, about 6,000 years ago, the ox became our first beast of burden. A millennium later, the first wild horses were captured from the steppes of Central Asia and bred for food. Again, it wasn't long before this animal's potential as a source of labour was discovered and (literally) harnessed. By the middle of the second millennium BCE, horses were central to almost every human activity, from agriculture to industry and from trade to war. The horse accompanied us well into the industrial era, and as late as the 1930s, horse-drawn carts were a common sight in industrialised cities (Kavar and Dovč 2008).

Prime Movers

Humankind's quest for energy began with tools that concentrated our own muscle power; continued with the use of fire for heat, light and cooking; later involved the cultivation of secure, high-energy foods; and eventually led us to harness the muscle power of larger animals. The most recent leap forward has been the invention of machines that can run without the involvement of human or animal muscles. Scientists refer to such machines, which convert a naturally occurring source of energy into mechanical power, as 'prime movers'.

Around the third century BCE, the power of running water was first harnessed by the ancient Greeks. Over the next millennium, waterwheel technology spread throughout the Mediterranean and to most of Asia and northern Europe. While the design and efficiency of these machines improved steadily over time, medieval watermills had a power output of only a few kilowatts, roughly equivalent to a modern hair dryer. Nevertheless, the waterwheel remained the most efficient pre-industrial prime mover and was a key factor in Europe's technical supremacy during the early stages of industrialization.

The second most important pre-industrial prime mover was the windmill. Windmills were first used in Persia around the tenth century. As the name suggests, these were used to mill grain, and later to pump water for irrigation (Smil 2008). Despite these and numerous other innovations, the way energy was used did not substantially change from prehistoric times to the eighteenth century. By 1800, people were still using animal muscle for work and transport, animal and vegetable fats for lighting, biomass for heating and methods of agriculture that had not greatly changed for millennia.

This all changed with the Industrial Revolution. What began in England, thanks to plentiful and easily accessible coal reserves, spread to France, Germany, Italy and the United States, and eventually reached

Table 1.2. *The power of different prime movers*

	Early use	1800s	1900s	2000s
Waterwheel	< 1 kW	200 kW	400 kW	
Windmill	< 1 kW	100 kW	400 kW	
Water turbine		5 kW	10,000 kW	1,000,000 kW
Steam engine	4 kW	100 kW	5,000 kW	
Steam turbine		10 kW	1,000 kW	2,000,000 kW
Gas turbine			100 kW	200,000 kW

Source: Smil (2008).

most of the world. New machines and tools allowed, for the first time, thermal energy to be converted into kinetic energy, driving a wide variety of machines. While many of the applications, such as mining and weaving, had been practiced for centuries or even millennia, the new machines allowed for great advances in scale, speed and efficiency. Just as horsepower improved early farmers' ability to work the land, the harnessing of thermal energy through the steam engine, and later the steam turbine, allowed a leap forward in manufacturing.

The invention of the steam engine set in motion a chain reaction of innovation and consumption that continues to this day. As people moved from villages to cities to work in factories, their habits of energy consumption changed. European and American populations exploded in the nineteenth centuries, and so too did the pressure on existing energy resources. Just as we face the challenge of diminishing fossil fuel resources today, societies of the nineteenth century had to find ways to replace wood and organic oils, up to then the principal fuels for heating, cooking and light.

The first country in the world to break its reliance on biomass fuel was England. Because of the twin demands of building (ships and houses) and fuel (for industry and domestic needs), most of the great English forests had already been cut down by the mid-1500s. The only way to avoid economic collapse was to find an alternative fuel. That alternative was coal. Coal was not exactly a new discovery; it had been used as early as 200 BCE by the Chinese and in Europe since Roman times (Smil 2008, Smith 1997). However, until the emergence of the British coal-mining industry, coal was extracted only from outcrops or shallow seams. New inventions such as steam-driven pumps allowed for larger and deeper mines. From the second half of the sixteenth century onwards, coal was mined extensively in England and Scotland, and by 1700 it had replaced wood as the main heating fuel. This early adoption of coal, and the resulting head start in terms of extraction methods and technologies, made Britain the cradle of the Industrial Revolution.

The development of railways allowed coal to be transported cheaply over long distances. Railways are an excellent example of how technical innovation exercised both a pull and a push effect on energy consumption: large amounts of coal were needed to produce steel for the railways, and the railways in turn allowed coal to be transported in bulk to the steelworks.

The Industrial Revolution changed not only the amount of energy consumed but also the way it was used. While coal sufficed initially as a fuel for locomotives, the internal combustion engine required high-energy liquid fuels. This led to the discovery of our most versatile fuel source to date: mineral oil. The ancient Chinese, Babylonians, Persians, Greeks and Romans had known about and used petroleum as a fuel for lighting and heating. However, it was not until the nineteenth century that mineral oil was used on an industrial scale. This was partly because it was difficult to extract, but also because it is dirty and inefficient in its raw state.

In preindustrial societies, most people were much less active at night. Candles and lamp oil were expensive and therefore used sparingly by all but the wealthy. Instead, people sat around fires at night, exchanging stories or performing stationary work such as mending clothes or tools. With the Industrial Revolution came a great migration into cities, rapid population growth, and the availability of much cheaper, factory-made goods. In cities, people were less inclined – and usually couldn't afford – to limit their daily work to the hours of natural light. The explosion in demand for lighting oil begat the whaling industry. Sperm whales yielded oils that burned far more cleanly and brightly than other animal fats did. By the time Herman Melville published *Moby-Dick* in 1851, the U.S. whaling fleet alone numbered 700 ships and was unloading 160,000 barrels of whale oil each year in the ports of New England (Smil 2008). As whale populations rapidly declined, coal gas and kerosene came to the rescue – of both the whales and the human consumers.

The world's first oil tycoon was neither a Texan cowboy nor an Arab sheik. Ignacy Łukasiewicz, a Polish pharmacist from the town of Gorlice on the fringes of the Austrian Empire, had experimented for several years with ways of distilling mineral oil. His breakthrough came in 1853 when the local hospital borrowed one of his kerosene lamps to conduct an emergency operation at night. Impressed by how brightly and cleanly the lamp burned, the hospital placed an order for further lamps and fuel. Łukasiewicz soon abandoned his pharmacy business to concentrate on the commercial application of his discovery. Within ten years he was not only mass-producing kerosene and lamps but was also the owner of several oil wells.

The first large commercial oil fields were tapped in the Caucasus region in the late nineteenth century. By the 1930s all of the world's leading economies were heavily dependent on oil, and by the middle of the century many of the world's biggest oil fields had been discovered. Primary among these were the immense fields of the Persian Gulf region, which hold, by present estimates, two-thirds of the world's reserves. The discovery of major oil fields was accompanied by the invention of new transportation technology, in much the same way as electricity generation and supply later spawned the development of myriad new electrical devices. The age of mass mobility had begun.

Demand for Liquid Fuels

The development of fossil-fuelled transportation removed one of the greatest limitations on land transport. Until the nineteenth century, raw materials, goods and people could only be transported over land using the muscle power of horses, oxen or camels. This made land transport costly and slow, and as a result only high-value goods, such as silk, precious stones and spices, were transported this way. The reliance on animal power for land transport extended well into the twentieth century. Railways initially took over the function of long-distance transport over land, but the horse was still needed to transport goods and people within the rapidly growing cities of Europe and North America. By the end of the nineteenth century, there were 300,000 working horses in London, one for every twenty people (Smil 2008). However, once the internal combustion engine reached mass production (by the 1920s in Europe and North America), horses quickly disappeared from the streets of Western cities.

We tend to think of our modern dependence on oil largely in terms of gasoline, yet the role of oil in modern industrial economies goes far deeper. As well as providing the basis of most transport fuels, oil is also the raw material for most fertilisers and pesticides, various chemicals, plastics, artificial fibres, lubricants, tar and asphalt. A trip to the supermarket by car relies on oil in many more ways than just the transportation fuel. First, there is the car itself, which, except for the chassis, engine and wheels, is largely made from oil-derived polymers. Second, the road on which we drive was probably constructed using asphalt. Finally, many of the clothing items, most of the packaging and much of the food we may buy at the store were produced with the direct or indirect involvement of mineral oil.

Most of us are aware of the importance of petroleum oil in satisfying the energy needs of modern civilisation, so it has become axiomatic to refer to our modern age as the age of oil. Yet it would be more correct to call it the fossil fuel age, since coal remains a much-utilised (and indeed growing) fuel for electricity production, and natural gas, the most recently harnessed of the fossil fuels, could soon rival oil in importance. A mixture of methane, butane, propane and other hydrocarbons, natural gas occurs either alongside mineral oil or dissolved within it. Like oil, natural gas has been known to humans for millennia, but it was even harder to utilise because of its form and volatility. The earliest known use of natural gas was in China during the Han dynasty (200 BCE), when it was siphoned from shallow underground pockets using bamboo tubing and used to boil seawater for salt production.

The first industrial-scale gaseous fuel was not natural gas, but town gas, a synthetic derivative of coal. Much of the street and domestic lighting in European and North American cities of the late nineteenth and early twentieth centuries was provided by town gas. Gasworks were an iconic feature of many industrialised towns and cities until the 1960s, by which time town gas had been largely replaced by electricity and natural gas. Three innovations were needed before natural gas could become a major household and industrial fuel: development of safe burners for mixing gas and air, wider high-pressure pipelines, and gas compression. Though a relative latecomer to the energy mix, natural gas has become the preferred fuel of the modern age for heating, cooking and electricity generation.

Energy at the Flick of a Switch

One could argue that we are still living through the Industrial Revolution, as most of our energy is still generated by burning fossil fuels. The main difference today is that we have added a new link to the energy conversion chain: electricity. The industrial generation of electricity represents an energy revolution in its own right. While the ancient Greeks had some understanding of electricity, as reflected in the origin of the term, it remained a scientific curiosity until the early nineteenth century. Thanks to the work of scientists such as Michael Faraday and Thomas Edison, electrical power was generated and harnessed for a variety of purposes.

Faraday led the way by discovering electrical induction; that a magnet moving within a copper coil will generate electrical current. This paved the way for the first electrical turbines, capable of converting mechanical to electrical energy. Edison's contribution to the development of electricity was even more profound. Like James Watt, Edison was both a scientist and a businessman. This gave him a strong incentive to develop machines for generating electricity and to provide a commercial system to transmit and distribute it. Edison built and operated the first power station in the United States, and he invented numerous devices capable of using electrical current, most famously the incandescent lightbulb.

The great advantage of electricity over combustion fuels is that it is rapidly and efficiently transportable and can be converted to other forms of energy (mechanical, thermal, light, etc.) at relatively high rates of efficiency. Moreover, it is clean, and can be made available instantaneously – literally 'at the flick of a switch'.

Energy at War

The Battle of Mons was one of the first engagements of the First World War. It began in late August 1914 when British cavalry happened upon their German counterparts on the French-Belgian border. The British riders chased the Germans for several kilometres before dismounting and engaging them in a gun battle.

Imagine muskets or swords in place of rifles, and a very similar battle could have taken place several centuries, even millennia, before. Yet just thirty years later the technology of warfare had been transformed, driven by enormous investment in military research through two world wars, the second of which ended with the discovery and use of a vastly more destructive weapon than any hitherto known. J. Robert Oppenheimer, one of the scientists who led the Manhattan Project, said of this new technology: "It has led us up those last few steps to the mountain pass; and beyond, there is a different country" (Rhodes 2010, p 3).

Solar radiation is the result of the fusion of hydrogen atoms in the sun's core, producing helium. However, the helium released has a slightly smaller mass than the sum of the hydrogen atoms. The 'missing' mass has been released as energy. This reaction was described in Einstein's famous formula $E = mc^2$. What this means is that the energy (E, expressed in joules) of any matter is equal to its mass (m, expressed in kilograms) multiplied by the speed of light (c) squared. Since the speed of light is immense (300,000 kilometres per second), the amount of energy theoretically contained in any kilogram of matter is similarly vast.

The discovery of nuclear energy opened up, as Oppenheimer foresaw, new horizons. Up to that time, most energy conversion involved combustion – first of biomass, then of fossil fuels. Nuclear energy represented a huge technological breakthrough: the ability to harness the most primal energy source of all – that of atoms and stars. Initially, many believed that this heralded an age of limitless energy. Lewis Strauss, chairman of the United States Atomic Energy Commission, claimed in a 1954 speech, "Our children will enjoy in their homes electrical energy too cheap to meter" (Smil 2010, p. 31). Strauss's claim was not as illusory as it may now appear. It was based not on the promise of nuclear fission, which relies on the relatively rare metal uranium, but on the belief that humans would one day harness the power of nuclear fusion, using the most abundant element on Earth, hydrogen.

The last great energy transformation of the twentieth century was the discovery, or rather rediscovery, of renewable energy. Until the Industrial Revolution, all the external energy sources used by humans – animal power, biomass, wind and water – were renewable. The rediscovery began with waterpower and the construction of large dams with turbines to produce electricity. In many countries hydropower played a major role in industrial development and urbanisation. Following the oil crises of the 1970s, there was also concern about energy independence and the future viability of fossil fuels. This greatly boosted research into energy alternatives. As a result, technologies such as wind turbines and photovoltaic (PV) solar panels, which had until then only been used for very specific and limited purposes (such as satellites), were developed for commercial use. By the late twentieth century, an additional impetus for renewable energy had emerged: concern about global warming and climate change.

2
Where Does Energy Come From?

2.1 Energy Commodities

Every form of energy that we currently use comes from the sun. The sun emits the light and heat that powers solar panels and water heaters, causes the air movements that drive wind turbines, replenishes the rivers that feed hydroelectric reservoirs and stimulates biofuel crops to grow, as it did the plants and algae whose fossilised remains form the coal, oil and gas in the Earth's crust. The sole exception to this rule is uranium, which did not so much come *from* as *with* the sun, having been present in the primordial nebula that gave rise to our solar system.

When we talk about energy sources, we are not describing their scientific origin as much as the form the different energy carriers take. That is why coal, wind and light are regarded as different energy sources. But because energy is big business, we also talk about energy commodities, meaning the different 'products' on the energy market (IEA 2005). Energy commodities can be divided into three main categories: heat, electricity and fuels.

While it is possible to obtain heat directly from the Earth and the sun, most of the heat we use in homes, offices and factories was obtained indirectly, usually by burning fuel. Sometimes, at a domestic scale, heat is generated using electricity; for example, in electric boilers or heat pumps. This method is, of course, highly inefficient as it represents a threefold energy conversion; fuel to heat, heat to electricity and electricity back to heat.

Electricity is the most versatile of all energy carriers. It is used to power a vast array of machines, from factory robots to domestic appliances. Its only limitation is storability, which is why we still don't have electric trucks or airplanes. It powers most of the conveniences

Table 2.1. *Primary and secondary fuels*

Primary fuel	Secondary fuel	Origin
Mineral oil	Petrol, kerosene, diesel	Fossil
Coal	Coke	Fossil
Wood	Charcoal	Renewable
Natural gas	Liquefied natural gas	Fossil
Bioethanol	ETBE	Renewable
Vegetable oil	Biodiesel	Renewable

of modern life, from electric lighting and food processing to televisions and computers. Electricity that is obtained directly from natural sources – such as hydro, wind, solar, tidal and wave power – is referred to as primary electricity, while electricity produced in nuclear or fossil-fuelled power plants by heating water to drive steam turbines is known as secondary electricity.

In many languages the word for fuel is closely related to combustion. After all, the ability to extract energy from fuels without burning them – through nuclear fission or fuel cells – is a recent discovery. Fuels that are sourced directly from nature, including wood, oil, coal, natural gas and uranium, are referred to as primary fuels. They may be used as they are or processed to produce secondary fuels (see Table 2.1). This is done either to facilitate transport (e.g., liquefied gas) or to produce derivatives with different properties (e.g., coke for metallurgy from coal, petrol from oil).

Combustion is a sequence of chemical reactions involving oxygen and a fuel, with energy released in the form of heat and light. Unless it is interrupted, a combustion reaction will continue spontaneously until all the chemical bonds are broken and the fuel is fully consumed.

2.2 Alternative, Sustainable and Renewable Energy

Before we dive into the detail of how energy is generated, it is well to clear up possible confusion about three similar but discrete terms. The term 'alternative energy' first came into common use in the 1970s during the oil crises and simply denoted alternatives to oil. At that time, the desire for such an alternative was driven more by political and economic considerations than by environmentalism or concerns about sustainability. Therefore, coal and natural gas, as well as nuclear, solar and wind power, were embraced as alternative energy sources. More recently, the term 'alternative energy' has been used to mean all alternatives to fossil fuels.

The term 'renewable energy' is more complex. Strictly speaking, fossil fuels are renewable since they are formed from dead biomass. However, the time span of their formation – millions of years – makes them non-renewable in practical terms. No matter how carefully or economically we use fossil fuels, they cannot be replaced at the rate of consumption. Therefore, a more practical definition of renewables might be "any energy storage reservoir which is being 'refilled' at rates comparable to that of extraction" (Sorensen 2004, 17). In the case of biomass fuels, we have the option of consuming them at the rate they are regenerated. This is the idea behind forest management; the biomass resource is cultivated in much the same way as an agricultural crop, maximising production while ensuring that the resource remains viable over time.

Solar and wind energy are, strictly speaking, not renewable, but rather inexhaustible, at least within a human time span. By the time the sun finally runs out of fuel (roughly 5 billion years from now), human beings will have long ceased to exist, at least in our present form. Solar energy is beyond our control, and we can neither deplete nor cultivate it. The amount of this energy we harness depends only on our will and ingenuity.

The terms 'renewable' and 'sustainable' are often used interchangeably, but this is also not quite right. Sustainability is a measure of the effects rather than the means of energy production. The term entered common use in the 1980s as more people became aware of the difficulties of reconciling economic growth with environmental protection. The influential Brundtland Report of the United Nations defined sustainable development in 1987 as "development that meets the needs of the present without compromising the ability of future generations to meet their own needs" (United Nations 1987, 41).

Energy that is renewable is not necessarily sustainable, and in recent years many have questioned the sustainability of some renewables, particularly biofuel production that competes with food, and wind farms in areas of scenic beauty.

2.3 Fuels: The Elixir of the Industrial Age

Anything that contains mass (in essence, anything we can see and touch) also contains energy. In a mass-to-energy conversion, mass physically disappears and is replaced by pure energy. According to Einstein's famous formula, $E = mc^2$, if we could fully convert matter into energy, a tiny pebble weighing a single gram would release 90,000 billion joules or 25 million kWh, enough to meet the electricity needs of a European town of 20,000 people for a year. If full mass-to-energy conversion were possible, all our energy problems would be solved.

But we're not there yet. While, in theory, all matter contains the same amount of energy, because of the specific chemical makeup of different molecules, some substances are more reticent about releasing their energy than others. Those that release it most readily – organic compounds consisting mostly of hydrogen and carbon (hydrocarbons) – are suitable as combustion fuels. Yet even these fuels release only a tiny fraction (less than 0.001 per cent) of their embodied energy. Nuclear fission, the reaction that comes closest to unlocking the potential in Einstein's formula, also only converts 0.5 per cent of the total mass of the uranium fuel into energy.

The most important characteristic of a fuel is what we call its 'energy density' (see Table 2.2). This is the amount of available energy in a given unit of a fuel. If we compare hydrogen and petrol, hydrogen has a higher energy density per unit mass than petrol, but a much lower energy density per unit volume, even in liquid form. This means that one kilo of hydrogen releases more energy than one kilo of petrol, but one litre of petrol releases more energy than a litre of hydrogen (see Figure 2.1).

Oil: The Slippery Slope to Fossil Fuel Dependency

When we think of oil fields, there's a tendency to imagine viscous underground lakes. However, the word 'petroleum', from the Greek *petra* (rock) and *elaia* (oil), betrays its true origin: as oily rock. Because this rock is buried deep within the Earth and is subject to enormous pressure, once we drill into the rock, the liquid is naturally pushed to the surface. Oil is a complex solution of different hydrocarbons; long molecules made of a scaffold of carbon atoms to which hydrogen atoms are bound (see Figure 2.3). The longer the molecule, the heavier and more viscous the substance. Substances with shorter molecules, such as methane, butane and propane, are gaseous at normal temperature and pressure (see Figure 2.2).

The eighteenth-century Russian scientist Mikhail Lomonosov was the first to argue that petroleum is biological in origin. Since then, scientists have confirmed that fossil fuels develop in the absence of

Table 2.2. *Energy density of different fuels in Joules per kilogram of mass, plus the main uses of each fuel and the type of energy involved*

Storage material	Energy per kg	Direct uses	Energy type
Enriched uranium (3.5% U-235)	3,500,000 MJ	Electric power plants (fast breeder reactor)	Nuclear
Natural uranium (0.7% U-235)	440,000 MJ	Electric power plants (light water reactor)	Nuclear
Hydrogen (at 700 bar)	130 MJ	Experimental automotive engines	Chemical
Natural gas (pipeline pressure)	53 MJ	Heating, electric power plants	Chemical
Petrol (gasoline)	45 MJ	Automotive engines	Chemical
Diesel	44 MJ	Automotive engines	Chemical
Jet fuel, kerosene	43 MJ	Jet engines	Chemical
Crude oil	42 MJ		Chemical
Biodiesel	37 MJ	Automotive engines	Chemical
Fat (animal/vegetable)	36 MJ	Human/animal nutrition	Chemical
Coal anthracites	32 MJ	Electric power plants, home heating	Chemical
Ethanol	30 MJ	Automotive engines	Chemical
Coal lignites	24 MJ	Electric power plants, home heating	Chemical
Carbohydrates	17 MJ	Food intake	Chemical
Wood (air-dried)	16 MJ	Heating, outdoor cooking	Chemical
Dry dung	10 MJ	Heating, outdoor cooking	Chemical
Municipal solid waste	9–16 MJ	Electric power plants	Chemical
Muscles	7 MJ	Food intake	Chemical
Gunpowder	3 MJ	Explosives	Chemical

oxygen from decomposed organic matter, forming sedimentary rocks rich in organic content. Organic matter is transformed, depending on the temperature and pressure, into different types of hydrocarbons, and that is why we find gaseous, oily or mixed reservoirs. While oil and natural gas were formed from plants, algae and animals that died

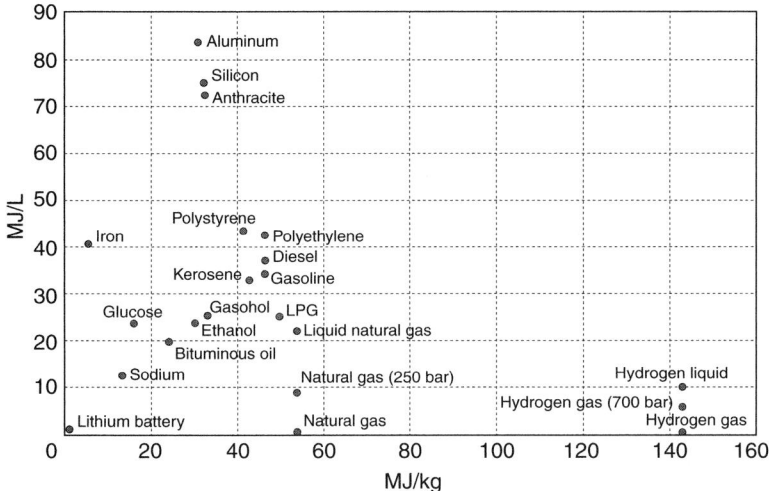

Figure 2.1. Energy density in terms of mass and volume of different materials and fuels. Hydrogen has a very high energy density per kilogram of mass, but being a gas, even at 700 bar of pressure its volume is large and it has a low energy density by volume. Batteries have low energy density both by volume and mass. Liquid fuels offer the best compromise.

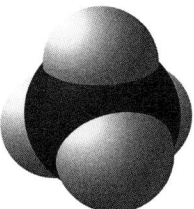

Figure 2.2. A 3-D model of methane and propane molecules, the two main components of natural gas and liquefied petroleum gas (LPG). They are composed of carbon and hydrogen atoms (hydrocarbons). Molecules with 1 to 4 atoms of carbon in each molecule are, respectively, called methane, ethane, propane and butane.

and sank to the bottom of primordial seas, coal was formed on land from decaying forest soils (see Figure 2.4).

Mineral oil came into use as a transport fuel following the invention of the internal combustion engine in the late nineteenth century. Then, in the decade preceding the First World War, oil began to occupy the central role it still holds in the transport economy. First, shipping

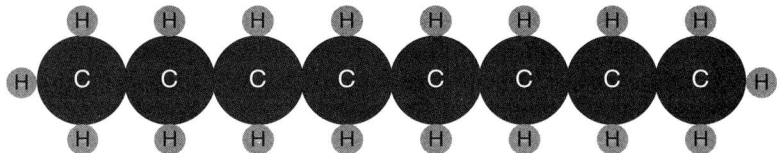

Figure 2.3. A 2-D model of octane, a molecule common in petroleum. It has 8 carbon atoms in a chain attached to hydrogen atoms. Of all hydrocarbons, octane makes the best motor fuel. Therefore, any motor fuel that burns as well as pure octane is termed 100 octane. Most fuels are between 85 and 90 octane.

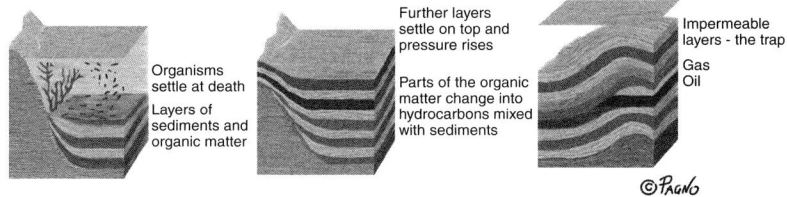

Figure 2.4. How petroleum and natural gas are formed. Algae and animals died and their bodies sank to the ocean floor. Over thousands of years, they were covered by layers of sediments (silt and sand). Over millions of years, further layers of sediment compress the organic matter, increasing pressure and turning the decaying organic matter into oil and gas. Fundamental to the formation of a deposit is the occurrence of an impermeable layer of sediment, which acts as an airlock.

switched from coal to crude oil because it ensured better flotation, higher speeds and longer periods between refuelling. Next, mass production made cars affordable for a mass market. Then, the railways switched from coal to diesel engines, and, finally, from the 1950s, kerosene-fuelled mass air travel completed oil's inexorable conquest of passenger and freight transportation.

Oil had three great advantages over coal: it is energy-dense, easy to transport by ship or pipeline, and was, at that time, cheap and abundant. Its dominance over other fuels was consolidated thanks to a steady decrease in price that began in the 1920s and only ended during the oil crises of the 1970s. Since then, a few other snags have become apparent: like other fossil fuels, oil is responsible for high greenhouse gas (GHG) emissions, and, more than any other fossil fuel, oil implies political and social instability, since nearly two-thirds

Figure 2.5. An oil field near Bakersfield, California.
Source: Gian Andrea Pagnoni.

of the world's reserves are concentrated in just six countries, five of them in the Middle East.

Natural Gas: The Lesser of Three Evils

Natural gas is a naturally occurring gaseous mixture that consists mainly of methane. Unlike coal or oil, methane does not take millions, thousands or even tens of years to form. It can form in a matter of hours as organic matter is digested in the gut of an animal. Methane is odourless, and it is the sulphur and other gases produced in the animal's gut that produce the characteristic smell. Modern biogas plants produce methane for commercial use by replicating the conditions inside a cow's gut. Natural gas, however, is a fossil fuel, formed, like oil, millions of years ago, and tends to be found alongside oil and coal deposits.

Greek legend has it that a goatherd first discovered a natural gas flare, the 'eternal flame', on Mount Parnassus. Believing this fire to be of divine origin, the ancient Greeks built a temple on the site. It was here that the priestess known as the Oracle of Delphi issued prophecies inspired by the flame. Despite such auspicious beginnings, the value of natural gas as an energy source was largely overlooked until recently. In fact, in the early years of oil exploration it was regarded more as a nuisance than a boon, a dangerous accessory to oil that was burned off to prevent explosions. It wasn't until the technology to safely extract, store and distribute natural gas was developed that its value could be appreciated.

These days, natural gas is as highly prized as oil, mainly as a fuel for electricity generation and heating. Natural gas is transported through high-pressure pipelines or by ship in the form of liquefied natural gas (LNG). Its main advantages over coal and oil are its cleanliness (it releases only carbon dioxide and water on combustion) and relative ease of transportation. The United States was the world's largest producer of natural gas until 1982, when it was surpassed by the USSR. In the last decade, the United States has been catching up again, thanks to a recent boom in shale gas.

Coal: The Mainstay of the Industrial Age

In the West most people instinctively associate coal with a bygone age: smokestacks, sooty factory towns, workers toiling in pits and blazing ironworks. There is a general assumption that we have moved on to cleaner and more sophisticated fuels. Yet the coal age never really ended. It is true that the energy mix diversified during the twentieth century as first oil, then natural gas, nuclear and renewable energies became commercially competitive. Yet coal has hung doggedly in there and remains the principal fuel of industrial production. It is just that production has become invisible for many in the West, as it now predominantly takes place in the East. The global economy would grind to a halt without coal, as it is by far the most important source of energy for electricity generation. Moreover, in the short term at least, its importance is likely to grow. As oil and gas supplies dwindle, we are likely to see a renaissance of coal precede a renaissance of renewables.

Coal is a sedimentary rock, composed primarily of carbon, as well as other minerals, including sulphur. It is formed through a process known as carbonization: leaves and branches fall and are compressed, first forming peat, then various qualities of coal, then graphite, and finally, under the right conditions, diamonds. Peat takes several thousand years to form, while anthracite, the most carbon- and energy-dense of the coal varieties, is formed over many millions of years. Graphite and diamonds are made of pure carbon and are not used as fuels. This is not just because of their rarity or usefulness for

Table 2.3. *Energy content of different fossil fuels compared with air-dried wood*

Fuel	Energy content
Poor brown coal or peat	> 8 MJ/kg
Brown coal (lignites)	> 14.7 MJ/kg
Air-dried wood	ca. 15 MJ/kg
Subbituminous coal (anthracites)	> 22.1 MJ/kg
Bituminous coal (anthracites)	> 26.7 MJ/kg
Crude oil	42–44 MJ/kg

other applications, but because their molecular bonds are too tight to allow for easy combustion.

Unconventional Oil and Gas: Scraping the Barrel

Conventional oil and gas – those reserves that are extracted using long-standing technologies such as vertical wells – are the 'low-hanging fruit' of the fossil fuel sector. Obviously, a term such as 'conventional' is subject to change: what is unconventional today may well be conventional sometime in the future, as new extraction, refining and transportation methods are developed. There are currently four main types of unconventional oil and gas: tar sands, shale gas, shale oil and coal bed methane.

Tar sands are a kind of immature petroleum, in which the organic matter has not yet been transformed into oil by heat and pressure. These sandy sediments contain bitumen, a type of petroleum that is too viscous to transport by pipeline without dilution. Shale gas and oil are trapped in shale rock that is too densely compacted to allow extraction using conventional drilling technology. Coal bed methane (CBM) is found in coal seams that are too deep or of too poor a quality for commercial mining.

Reserves of unconventional oil and gas are both immense and widely distributed, but because of the greater costs and additional technology required to extract them, many are not yet competitive with conventional sources. However, a steadily rising market price for crude oil, growing uncertainty about the global supply, and rapidly growing demand from Asia have spurred on the unconventional oil and gas sector. Thanks to shale gas fracking in the United

States and bitumen melted out of the Alberta tar sands, the United States is poised to overtake Russia as the world's largest gas producer, and Canada has become the biggest exporter of oil to its southern neighbour (Biello 2012b).

Nuclear Fuels

There's more to heavy metal than long hair and shrieking guitar solos. In chemistry, the term refers to metallic elements that have a large number of protons in their nucleus. The heaviest of these, uranium, has so many protons that it naturally sheds neutrons. In a controlled environment, this property can be harnessed to generate electricity.

Like crude oil, uranium in its naturally occurring state is not of much use as a fuel. It occurs as an ore (most commonly uraninite, also known as pitchblende), from which uranium fuel is extracted. Though generally associated with the modern age, uranium has been known and used for centuries, mainly as a colourant for stained glass. It was also used in early photography to produce tints and shades.

Most people are at least vaguely familiar with the periodic table, the colourful graph representing all known chemical elements (see Figure 2.6). The table was devised in 1869 by the Russian chemist Dimitri Mendeleev to arrange the different elements based on certain characteristics, such as mass, size and chemical properties. The elements close to the bottom of the table are heavy, and some are unstable, as they contain a high number of protons that repel each other. An unstable atom releases particles and electromagnetic waves in a process known as radioactive decay, thereby transforming the atom into a different element.

Uranium, like all elements, occurs in different isotopes. Isotopes are forms of the same element containing different numbers of neutrons. By far the most common uranium isotope found on Earth is uranium-238. The number refers to the fact that the nucleus of this isotope contains 92 protons and 146 neutrons (92 + 146 = 238). The radioactivity level of uranium-238 is very low, and it is hazardous because of its toxicity rather than its radioactivity. Uranium-238 is therefore not suitable as a fuel for nuclear reaction.

A far rarer isotope (representing less than 1 per cent of naturally occurring uranium) is uranium-235, which is sufficiently radioactive to sustain a chain reaction. Uranium-235 contains 92 protons and 143 neutrons. Scientists found a way to extract uranium-235 from ore in a process known as enrichment. Gasified uranium is spun at very

Key:

element name **atomic number** **symbol** atomic weight (mean relative mass)

hydrogen 1 **H** 1.0079																		helium 2 **He** 4.0026
lithium 3 **Li** 6.941	beryllium 4 **Be** 9.0122												boron 5 **B** 10.811	carbon 6 **C** 12.011	nitrogen 7 **N** 14.007	oxygen 8 **O** 15.999	fluorine 9 **F** 18.998	neon 10 **Ne** 20.180
sodium 11 **Na** 22.990	magnesium 12 **Mg** 24.305												aluminium 13 **Al** 26.982	silicon 14 **Si** 28.086	phosphorus 15 **P** 30.974	sulfur 16 **S** 32.065	chlorine 17 **Cl** 35.453	argon 18 **Ar** 39.948
potassium 19 **K** 39.098	calcium 20 **Ca** 40.078		scandium 21 **Sc** 44.956	titanium 22 **Ti** 47.867	vanadium 23 **V** 50.942	chromium 24 **Cr** 51.996	manganese 25 **Mn** 54.938	iron 26 **Fe** 55.845	cobalt 27 **Co** 58.933	nickel 28 **Ni** 58.693	copper 29 **Cu** 63.546	zinc 30 **Zn** 65.38	gallium 31 **Ga** 69.723	germanium 32 **Ge** 72.61	arsenic 33 **As** 74.922	selenium 34 **Se** 78.96	bromine 35 **Br** 79.904	krypton 36 **Kr** 83.80
rubidium 37 **Rb** 85.468	strontium 38 **Sr** 87.62		yttrium 39 **Y** 88.906	zirconium 40 **Zr** 91.224	niobium 41 **Nb** 92.906	molybdenum 42 **Mo** 95.96	technetium 43 **Tc** [98]	ruthenium 44 **Ru** 101.07	rhodium 45 **Rh** 102.91	palladium 46 **Pd** 106.42	silver 47 **Ag** 107.87	cadmium 48 **Cd** 112.41	indium 49 **In** 114.82	tin 50 **Sn** 118.71	antimony 51 **Sb** 121.76	tellurium 52 **Te** 127.60	iodine 53 **I** 126.90	xenon 54 **Xe** 131.29
caesium 55 **Cs** 132.91	barium 56 **Ba** 137.33	57-70 *	lutetium 71 **Lu** 174.97	hafnium 72 **Hf** 178.49	tantalum 73 **Ta** 180.95	tungsten 74 **W** 183.84	rhenium 75 **Re** 186.21	osmium 76 **Os** 190.23	iridium 77 **Ir** 192.22	platinum 78 **Pt** 195.08	gold 79 **Au** 196.97	mercury 80 **Hg** 200.59	thallium 81 **Tl** 204.38	lead 82 **Pb** 207.2	bismuth 83 **Bi** 208.98	polonium 84 **Po** [209]	astatine 85 **At** [210]	radon 86 **Rn** [222]
francium 87 **Fr** [223]	radium 88 **Ra** [226]	89-102 **	lawrencium 103 **Lr** [262]	rutherfordium 104 **Rf** [267]	dubnium 105 **Db** [268]	seaborgium 106 **Sg** [271]	bohrium 107 **Bh** [272]	hassium 108 **Hs** [270]	meitnerium 109 **Mt** [276]	darmstadtium 110 **Ds** [281]	roentgenium 111 **Rg** [280]	ununbium 112 **Uub** [285]		ununquadium 114 **Uuq** [289]				

*lanthanoids	lanthanum 57 **La** 138.91	cerium 58 **Ce** 140.12	praseodymium 59 **Pr** 140.91	neodymium 60 **Nd** 144.24	promethium 61 **Pm** [145]	samarium 62 **Sm** 150.36	europium 63 **Eu** 151.96	gadolinium 64 **Gd** 157.25	terbium 65 **Tb** 158.93	dysprosium 66 **Dy** 162.50	holmium 67 **Ho** 164.93	erbium 68 **Er** 167.26	thulium 69 **Tm** 168.93	ytterbium 70 **Yb** 173.06
**actinoids	actinium 89 **Ac** [227]	thorium 90 **Th** 232.04	protactinium 91 **Pa** 231.04	uranium 92 **U** 238.03	neptunium 93 **Np** [237]	plutonium 94 **Pu** [244]	americium 95 **Am** [243]	curium 96 **Cm** [247]	berkelium 97 **Bk** [247]	californium 98 **Cf** [251]	einsteinium 99 **Es** [252]	fermium 100 **Fm** [257]	mendelevium 101 **Md** [258]	nobelium 102 **No** [259]

Figure 2.6. The periodic table of elements. An element is a chemical substance made up of just one type of atom. Every atom contains protons, neutrons and electrons. The number above the symbol denotes number of protons in the atom (this is known as the atomic number). The number below the symbol denotes the combined weight of the protons and neutrons. Electrons are so tiny that they have little influence on mass.

high speed in a centrifuge. Like the spin cycle of a washing machine, which draws water out of fabric, the centrifuge draws the heavier U-238 atoms to the outside, leaving a higher proportion of U-235 at the centre. Low-enriched uranium, with around 5 per cent uranium-235, is the most common fuel in modern reactors. One kilogram of this fuel can produce the same amount of energy as 2,000 tonnes of coal in a conventional power plant. Nuclear weapons require a far higher level of enrichment (more than 80 per cent U-235). The difficulty and expense of producing highly enriched uranium is one of the reasons why so few countries possess nuclear weapons.

Biofuels

Initially hailed by environmentalists and industrialists alike, the one renewable source that seemed to please everyone, biofuels have fallen from grace in recent years. Since the 2007–2008 global food crisis, when food prices increased dramatically, partly in response to a growing demand for ethanol, biofuels have been seen by many as a way of diverting food from the plates of the world's poorest to the gas tanks of its richest. Even the definition of biofuels is in dispute. Strictly speaking, they include all fuel sources derived from biomass; namely, wood, liquid fuels (ethanol and biodiesel) and methane gas obtained from biological sources. However, because biofuel production has been subsidised by many countries as a means of reducing carbon dioxide emissions and dependency on fossil fuels, the term has become important from legal, political and economic perspectives. In this sense the term 'biofuel' is understood to mean the liquid fuels derived from plant sources and used as a replacement for petrol and diesel in transport. The most common of these are bioethanol, mainly produced by fermenting sugarcane (in Brazil) and maize (in North America), and biodiesel, mainly produced by refining rapeseed and sunflower oils in Europe.

Biofuels can be used in much the same ways as fossil fuels: for transport, household heating, and in power plants. The main difference between biofuels and fossil fuels is that biofuels are renewable and the carbon dioxide they give off when burned is largely counterbalanced by that absorbed during photosynthesis as the plants grow.

2.4 Energy Storage

When we overeat, we take in more food energy than we expend through our daily activities. Our body automatically stores the surplus energy in the form of fat, a concentrated energy source (it has double the energy density of carbohydrates) that sustains animals over longer periods without food. The way we store fat also has a lot to do with our external environment. People living in cold and temperate climatic zones adapted by building up stores of subcutaneous fat during summer and autumn. Those from tropical and hot climates didn't require this annual energy buildup.

Civilisation began with humans' ability to store energy outside their bodies. By storing food and fuel, we were able to maintain a constant supply of energy in a changing environment. Over the last 200 years our capacity to store energy has, on the one hand, been greatly enhanced and, on the other, been retarded by the increased use of fossil fuels. Nature has already done the hard work of creating these energy stores, and we have only needed to find ways of extracting and using the energy. Therefore, we have been able to largely ignore the issue of large-scale energy storage until now. Even our most sophisticated energy storage devices – electrical batteries – remained largely unchanged until the late twentieth century. In fact, the earliest known model of rechargeable battery – the lead-acid battery invented by Gaston Planté in 1859 – is still used in modern automobiles.

Small-Scale Storage: Batteries

The first batteries bore little resemblance to the neat little slabs that slot into mobile phones and other portable devices. In fact, the term 'battery' came into being because the earliest such device, a row of Leyden jars used to store static electricity, reminded Benjamin Franklin of a battery of cannon. All batteries work by converting chemical energy into electrical energy. They consist of two parts, a cathode and an anode, each containing a different chemical compound. The anode releases electrons, which are attracted towards the cathode, creating an electrical current. In a lead-acid battery, the anode is made of metallic lead and the cathode of lead dioxide. The acid merely acts as a medium through which current can flow. Dry batteries, such as those used in laptop computers and mobile phones, generally use metals and oxides as anodes and cathodes.

Great strides have been made in electricity storage in the last thirty years, as battery technology has adapted to suit an ever-increasing array of electronic devices. Yet batteries remain capable of storing only relatively small amounts of energy. In 2012 the State Grid Corporation of China, in collaboration with the electric car manufacturer BYD, unveiled the world's largest battery, capable of storing 36 megawatt-hours of electricity. This facility is larger than a football field and cost $500 million (U.S. million) to build. When we consider that a small 500-megawatt coal-fired power plant costs about the same amount to build, yet produces 36 megawatt-hours every four minutes, it becomes clear that electrical energy storage is not even close to industrial scale yet.

This is a particularly vexing issue for the electric car industry. Although the electric car is widely regarded as the vehicle of the future, it has actually been around just as long as its gas-guzzling counterpart. In fact, many of the early speed records were set by electric cars, prompting the *New York Times* in November 1911 to extol their virtues. At present, the best small electric vehicles can only run for about 200 to 300 kilometres between charges, and charging takes several hours. For electric cars to replace conventional cars, another great leap in battery technology will be needed (see Chapter 7).

Large-Scale Storage

So, if batteries are not capable of storing energy at an industrial scale, what other options do we have? At present, elevated reservoirs of water are the most economically viable way of storing large amounts of energy. This method is already used in hydroelectric power plants, where huge quantities of water are stored in reservoirs. The most difficult part of this operation is accomplished by the sun, which raises the water from the oceans to the mountains through evaporation and precipitation. With pumped-storage hydroelectricity (PSH), humans assume the role of the sun, pumping water uphill to build up a reservoir of potential energy (see Figure 2.7).

This would appear at first to be a Sisyphean task, and indeed it does involve a net loss of energy. It makes sense, however, in the context of power generation. Fossil fuel and nuclear power stations produce a constant supply of electricity, yet demand varies considerably between day and night. It would not be economical to shut down all or part of a power station at night, so instead generation companies can use pumped storage to make use of the surplus electricity generated at night. Large-scale storage also makes sense in the case of wind and solar power plants: wind blows intermittently and the sun does not shine at night. Storage is therefore required to compensate for calm periods and for nights and cloudy days.

Figure 2.7. Seneca Pumped Storage Generating Station above Kinzua Dam on the Allegheny River, Pennsylvania, United States.
Source: U.S. Army Corps of Engineers/Margaret Luzier.

Like water, air can also be stored for later release to drive turbines. In this case, air is pumped at high pressure into underground caverns, abandoned mines, or large tanks. Within the current energy economy, large-scale water and compressed air storage is not viable. However, in a future energy economy where renewables play a greater role, methods such as these can help to compensate for the intermittency of wind and solar energy.

Heat Storage

Since heat is the vibration of scattered and unordered molecules, storing it is a bit like herding cats; its natural tendency, consistent with the second principle of thermodynamics, is to disperse. There are two main approaches to storing heat. The first involves strategies such as insulation and passive-house technologies to retain heat or minimise its loss in buildings. The second approach involves storing concentrated heat for later use as a primary energy source. It is this second approach that concerns us here. We will deal with insulation when we talk about energy efficiency.

Heat is a by-product of every energy conversion. The main challenge in terms of energy efficiency is to retain and utilize this heat. A number of methods have been developed to either store heat for short periods or to distribute it quickly to local homes and offices. The most effective method of storing heat is that used by some concentrated solar power (CSP) plants. During the day part of the concentrated solar heat is used to melt salts, which are then stored at temperatures of up to 300 degrees Celsius in insulated tanks. At night or on cloudy days, this stored heat can then be used to drive steam turbines, thereby ensuring around-the-clock electricity generation.

Hydrogen

Like Coleridge's ancient mariner, lost at sea and lamenting the presence of "water, water, everywhere, nor any drop to drink" (Coleridge 1857, 13), we find ourselves surrounded by a profusion of the perfect fuel source, but not in a form we can use. Water, one of the most common substances on Earth, contains two atoms of hydrogen and one of oxygen in each molecule. On Earth, hydrogen is therefore mostly associated with water, as its name suggests (from the Greek *hydro*, meaning water, and *genes*, meaning creator). Beyond our planet, hydrogen gas is greatly abundant, fuelling our sun and comprising an estimated three-quarters of the entire mass of the universe.

Hydrogen seems like the answer to all of our energy problems: it burns extremely well in a clean reaction, releasing only water vapour[1]; it can be used in fuel cells to produce electricity; and perhaps in the future we will be able to use it in nuclear fusion reactors. As if that weren't enough, hydrogen is very easy to produce through electrolysis: an electrical current is passed through water, breaking the bonds of the water molecule and releasing hydrogen and oxygen.

Though the world's first internal combustion engine, developed in 1804 by the Franco-Swiss inventor François Isaac de Rivaz, used hydrogen, this was soon abandoned as an automotive fuel in favour of petrol and diesel. Since then, hydrogen engines have been confined either to the experimental sections of auto shows or to the aerospace industry (the space shuttles burned hydrogen as propulsion fuel). In fuel cells, hydrogen and oxygen are used as the anode and cathode of a battery, between which an electron flow occurs.

---

[1] Hydrogen easily reacts with oxygen in a combustion reaction where energy and water is released: $2H_2 + O_2 \rightarrow 2H_2O + energy$.

Figure 2.8. A hydrogen-powered car at a filling station. Safety concerns still exist, but the main hurdle to widespread use of hydrogen is economic.
Source: U.S. Navy.

If hydrogen is such a plentiful, versatile and clean fuel, why has it not already replaced fossil fuels? There are two main reasons. First, hydrogen is extremely difficult to store and it ignites easily, as was dramatically demonstrated by the Hindenburg airship disaster in 1937. The second, and arguably greater, obstacle to a future hydrogen age lies with economics rather than technology. Most of our fuels are a gift of time and nature; we have simply learned to refine them. Hydrogen is different, as it must first be manufactured using other energy sources. Hydrogen is plentiful but is locked in molecular bonds (e.g., in water or methane) that take considerable energy to break. That energy has to be generated by other means, in most cases by burning fossil fuels. This entails a net loss of energy and the emission of pollutants.

At an industrial scale, hydrogen is mainly produced from methane (CH_4), in a process known as reforming, where the hydrogen atoms are unlocked from natural gas molecules. If the current shale gas boom in North America continues, a plentiful supply of methane may lower the cost of hydrogen to a level where it is competitive with conventional fuels (Irfan and ClimateWire 2013).

At present, the value of hydrogen lies in its energy density and portability. Provided we can find ways to safely transport it in large quantities, hydrogen can provide a highly effective means of energy storage. That is why the future of hydrogen use will largely depend on the future of renewable energy. Unlike fossil or nuclear energies, renewables by their nature require accompanying storage strategies.

Where wind and solar technologies account for just a tiny percentage of the current electricity mix in most countries, storage is largely irrelevant. However, if we imagine a future where the wind and the sun are our main sources of energy, we will need a variety of storage approaches to get us through the doldrums and sunless days.

3
How Much Energy Is Enough?

3.1 Scales of Consumption: From Mobile Phones to Space Shuttles

On my first trip to New York, I felt a certain déjà vu. Everything seemed familiar, from the crystalline office towers to the murky Hudson, not to mention landmarks such as the Statue of Liberty and the Brooklyn Bridge. Looking out through the airplane window, I couldn't help thinking about the amount of water and waste, goods and garbage, fuels and fumes constantly flowing through the city. Later, when I found myself absorbed by the rushing current of humanity that is daytime Manhattan, I witnessed up close a strange marriage of entropy and order, as millions of people circulated in an area one-fifth the size of Malta. For the first time, I sensed that large cities lose the human scale, and that what you see from the plane is not that different from what you see on the streets: a flow of energy.

New York relinquished the title of world's largest city a few decades ago, and its tallest towers are dwarfed by more recent constructions in Asia, yet it remains a symbol both of the achievements and confines of urban development. An uninitiated visitor to New York, Tokyo, São Paulo, or any of the mega-cities of the twenty-first century, might ask, how is it possible that it all keeps going without ever running out of steam?

The answer is, thanks to gargantuan machines, the power plants tucked away out of sight, just beyond the suburbs of these great cities. But what if, for some reason, those machines suddenly stopped working and we were suddenly cast back into an age when human muscle power was the principal source of energy? How much manpower would it take to charge a mobile phone or to run a computer? What would it take to keep the lights on in Times Square? We are so used

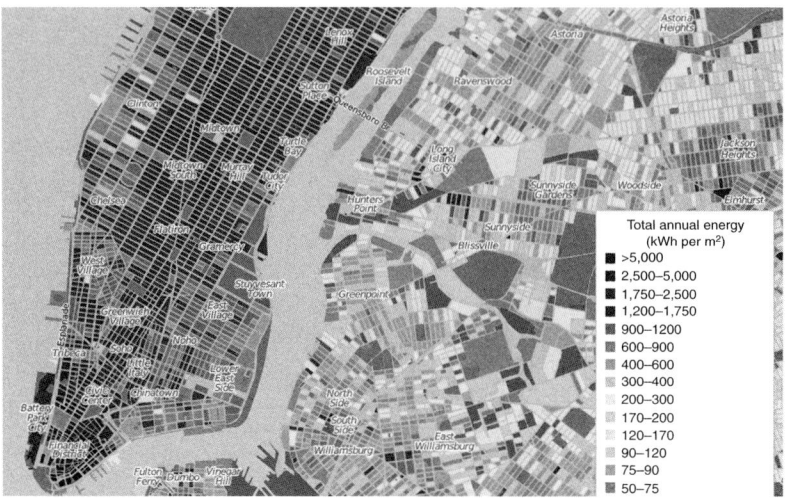

Figure 3.1. Energy consumption per square meter in New York City. Because of the heavy concentration of homes, offices, hotels, and shops in the most built-up parts of the city, the energy requirements per square metre of land are extremely high, up to five times the solar energy that strikes the ground.
Source: Howard et al. (2012).

to relying on external sources of energy that these questions seem absurd, and yet to understand how we consume energy, we must first return to the human scale.

Energy for Human Metabolism

For most of history, human energy needs consisted of food and fuel for heating and cooking. It was only quite recently that we started to use energy for a wider variety of applications. Now, in the early twenty-first century, we have reached a point where food represents just 1 per cent of the energy consumed by people living in wealthy industrialised countries. The remainder is used in transportation, by electrical appliances, for lighting and heating, and to power industrial machinery.

Most people are familiar with the term 'calorie' as the unit used to measure food energy.[1] Even though the joule is the standard unit

---

[1] A calorie is defined as the amount of energy needed to raise the temperature of 1 millilitre of water by one degree Celsius. One calorie (cal) is about 4.2 joules.

of energy, calories are often used to describe the amount of energy in food. When we talk about the number of 'calories' in a piece of cake or a serving of pasta, we are actually talking about kilocalories (kcal or Cal or 1,000 cal), yet the word 'calorie' has persisted in popular usage.

Our bodily energy needs depend not only on our level of activity but also on our age and size, as well as on environmental conditions such as the temperature of our surroundings, which determines whether we need to expend energy to warm or cool our bodies (see Table 3.2). An infant needs about 700 kilocalories (roughly 3,000 kilojoules) per day, while a very active adult man may consume up to 4,000 kilocalories (more than 15,000 kilojoules). Of course, energy is not the only thing we get from food. Food also provides the raw materials to build our cells. Because of this dual requirement, our diet needs to be varied, including foodstuffs such as fats and oils that are very high in energy, and others, such as vegetable fibres, that are low in energy but are vital to the maintenance and repair of tissues (see Table 3.1).

All of the body's organs need energy in order to function, and even while sleeping we consume quite a lot of energy. This minimum expenditure of energy just to stay alive is known as the basal metabolic rate (BMR). The energy expended in sedentary activities, such as sitting at a computer and writing these words, is about 1.5 times BMR. More physically strenuous activities obviously require more energy, up to a few times the BMR (see Table 3.3).

While sleeping, an adult male consumes energy at a rate of about 40 watts. At the other end of the human energy spectrum, an Olympic sprinter achieves a power output of about 2,000 watts, but only in short bursts. An adult male can maintain an energy output of about 80 watts for several hours – enough, for example, to cycle at a leisurely pace. For women, the equivalent output is about 60 watts (Smil 2008).

Let's compare this to the output of the machines that fill modern lives. A hair dryer, for example, consumes energy at a rate of more than 1,000 watts, equivalent to thirteen men riding bicycles connected to electric generators. A wealthy person might still be able to afford such a hair dryer, but what about powering a washing machine, an air-conditioning system, or a car? To replace the power output of the twenty solar panels that provide electricity for my home, I would need sixty grown men riding bicycles on my roof. I would need almost 200 men to power my car, and to produce the same amount of power as the world's largest power plant, the Three Gorges Dam in China,

Table 3.1. *Approximate energy content of common foodstuffs compared with fuels; data are expressed in the international unit measure of energy and in the other two common units, kilocalories and kilowatt-hours*

Food 100 gr	Fuel 100 gr	kJ	kcal	kWh
Watermelon		63	15	0.017
Tomato		63	15	0.017
Red wine (100 ml)		251	60	0.070
Milk (100 ml)		264	63	0.073
Banana		272	65	0.076
	Gunpowder	300	72	0.083
Potatoes		356	85	0.099
Lentils		377	90	0.105
Tuna fish (natural)		439	105	0.122
Chicken breast		460	110	0.128
Egg		544	130	0.151
Whole bread		962	230	0.267
	Hay	1,004	240	0.279
Vodka (100 ml)		1,025	245	0.285
	Coal (lignite)	1,213	290	0.337
Cheeseburger		1,234	295	0.343
White bread		1,255	300	0.349
Basmati rice		1,423	340	0.395
Cornflakes		1,485	355	0.413
	Wood	1,506	360	0.418
Cheddar cheese		1,632	390	0.453
Refined white sugar		1,632	390	0.453
Milk chocolate		2,259	540	0.628
	Coal (antracite)	2,700	645	0.750
	Charcoal	3,000	717	0.833
Butter		3,138	750	0.872
	Natural gas	3,300	789	0.917
	Biodiesel	3,330	796	0.925
	Diesel	3,640	870	1.011
Olive oil		3,772	901	1.048
	Crude oil	4,200	1,004	1.167
	Petrol	4,400	1,052	1.222
	Hydrogen	11,400	2,725	3.167

Table 3.2. *Daily energy requirements of infants and adults*

Energy requirements	Years	kg	Lifestyle	kJ/day	kcal/day
Infant	1	12	active	3,243	775
Child	5-6	30	active	6,139	1,467
Woman	30-60	55	sedentary	7,800	1,850
Woman	30-60	55	active	11,900	2,850
Man	30-60	70	sedentary	10,200	2,450
Man	30-60	70	active	15,400	3,700

Source: FAO 2001.

Table 3.3. *Energy required to conduct various activities, expressed as multiples of the basal metabolic rate (BMR), which in an average adult male is approximately 1,500 kcal/day (6,279 kJ/day)*

Lifestyle	BMR multiple
Basal metabolic rate	1
Sleeping	1
Sitting (eating, watching TV, etc.)	1.5
Standing, carrying light loads	2.2
Personal care (dressing, showering)	2.3
Household work	2.8
Walking at varying paces without a load	3.2
Agricultural work (planting, weeding, gathering)	4.1
Low-intensity aerobic exercise	4.2
Collecting water/wood	4.4

Source: FAO 2001.

I would need more than 250 million tireless cyclists. We take for granted access to quantities of energy unimaginable to pre-industrial societies (see Tables 3.4 and 3.5). Chances are that readers of this book enjoy a far greater energy budget than the kings and emperors of the ancient world.

Table 3.4. *The power in watts required to run various devices and the manpower that would be needed to produce the same amount of power*

Device	W	Working men
Mobile phone	3	0.04
CD player	25	0.31
LCD TV	100	1.3
Computer and monitor	150	2
Microwave oven	1,000	13
Hair dryer	1,200	15
Electric heater	2,500	31
Lawnmower	3,500	44
Average car at 80 km/h	15,000	188
Formula-1 car	470,000	5,875
Boeing 747 at takeoff	80,000,000	1,000,000
Nuclear submarine	190,000,000	2,375,000
Space shuttle at takeoff	12,000,000,000	150,000,000

Table 3.5. *The power in watts generated by various devices and plants, and the manpower that would be needed to generate the same amount of power*

Device/plant	W	Working men
Solar cell in pocket calculator	0.10	0.00
Small candle	3.00	0.04
Human	80	1.0
Typical PV solar panel	230	3
Ox	375	4.7
Wind turbine for a small boat	400	5
Horse	525	6.6
Typical household PV solar plant	3,000	38
Portable diesel generator	3,000	38
Watt's first steam engine	4,400	55
Typical household stove	15,000	188
Typical industrial rooftop PV plant	200,000	2,500
Typical biogas power plant	1,000,000	12,500
Typical wind turbine	2,000,000	25,000

Table 3.5 (continued)

Device/plant	W	Working men
World's largest PV plant (Huanghe Golmud Solar Park, China)	200,000,000	2,500,000
World's largest steam turbine	700,000,000	8,750,000
World's largest wind farm (Roscoe Wind Farm, USA)	781,000,000	9,762,500
Geothermal power plant (Larderello, Italy)	810,000,000	10,125,000
Typical coal power plant	1,000,000,000	12,500,000
World's largest nuclear power plant (Kashiwazaki-Kariwa, Japan)	8,200,000,000	102,500,000
World's largest power plant (Three Gorges Dam, China)	20,300,000,000	253,750,000

> **What Does It Take to Run ... a Fridge?**
>
> The fridge, an appliance that has become an indispensable part of most households in the industrialised world, exemplifies how abstract our relationship to energy has become. One hundred years ago, the main energy requirement for a household was fuel (usually wood or coal) to heat and cook. This was purchased or harvested, then stored in or next to the home. A typical farmhouse in northern Europe would include a storeroom filled with fuel to sustain the family throughout the winter. The energy used to run modern household appliances is abstract because it only appears as a figure on an electricity bill. We tend not to concern ourselves with the fuels used to generate that power. But if we were to run a domestic fridge ourselves, how much fuel would we need?
>
> A medium-sized fridge (200 litres) requires about 160 watts of power. If well insulated, it will need to run for only about four hours per day, or 1,460 hours per year. To calculate how much energy it consumes over a year, we multiple 160 watts (or 0.16 kilowatts) by the number of hours it runs for. This gives us 230 kilowatt hours or units of electricity consumed per year.
>
> The average chemical energy content of coal is 6,150 kilowatt-hours per ton or 6.15 kilowatt-hours per kilo. Although coal-fired power plants are quite efficient, they are still limited by the laws of thermodynamics. Therefore, only about 40 per cent of the chemical

(continued)

> energy in the coal is converted into electricity. So, a kilo of coal can generate 2.46 kilowatt-hours of power (0.4 x 6.15 kWh). To calculate how many kilos of coal we would need to run a modern fridge, we divide 230 kilowatt-hours by 2.46 kilowatt-hours. That gives us 95 kilograms of coal per year. If we now add a television, computers, lighting, heating, an electric cooker, a vacuum cleaner and other electrical appliances, we would need a large additional room just to store the fuel needed to power our appliances and heat our homes.
>
> Most electricity is generated in power plants using coal or gas. So, figuratively speaking, we all have our storehouse of coal, just not inside our homes. But what if we switched to renewables? A house provides a ready-made base for one of the most promising of small-scale renewable sources, solar photovoltaic (PV) panels. A modern domestic PV plant in a temperate climate generates an average of 150 kilowatt-hours per square metre per year. This means that a single 1.5 square meter PV panel can provide sufficient energy to run a modern fridge. Twenty panels would meet the electricity needs of a typical European household.

3.2 Energy Consumption by Sector

Energy supply involves the delivery of fuels, electricity, or heat from the point of production to that of consumption. It encompasses the extraction, transformation, generation, transmission, distribution, and storage of commodities (see Figure 3.2). Far more energy is produced than consumed because losses occur in conversion and delivery (see Figure 3.3).[2] Total world energy production is roughly 530 exajoules (EJ) – one million trillion joules (10^{18} joules; see Table 3.7) – while global final consumption after losses is 350 exajoules (BP 2011; IEA 2012a).

Energy consumption is one of the most accurate indicators of wealth. Two hundred years ago, a wealthy family in Europe or North America consumed about five times as much energy as a poor family in their own country (measured by the amount of food and fuel they used and the number of horses they kept), and about ten times as

---

[2] Total Final Consumption (TFC) is Total Primary Energy (TPE) less the quantities of energy required to transform primary sources, such as crude oil or wind, into forms suitable for end use, such as petrol or electricity.

Table 3.6. Energy (in joules) released by some phenomena, devices, and power plants

J	kWh	Nature of energy release
1	0.00000028	Kinetic energy produced as an apple (~100 grams) falls 1 metre
10	0.0000028	Flash of a typical pocket camera
300	0.000083	Energy of a lethal dose of X-rays
1,773	0.00049	Kinetic energy of M16 rifle bullet
17,0004.200	0.00470.0012	Energy released by the metabolism of 1 gram of carbohydrates or protein
38,00017,000	0.0110.0047	Energy released by the metabolism of 1 gram of fat
45,00038,000	0.0130.011	Energy released by the combustion of 1 gram of petrol
50,00045,000	0.0140.013	Kinetic energy of 1 gram of matter moving at 10 km/s
42,000,0003,600,000	11.71.0	Energy released by an Olympic swimmer during training
3,600,000,00042,000,000	1,00011.7	Annual solar radiation per m^2 in North Europe
5,000,000,0003,600,000,000	1,3891,000	Energy in an average lightning bolt
4,500,000,0005,000,000,000	1,2501,389	average annual energy usage of a standard refrigerator
9,000,000,0004,500,000,000	2,5001,250	Average solar radiation per m^2 in sunny deserts
18,000,000,0009,000,000,000	5,0002,500	Annual electric yield of a typical household PV plant
90,000,000,0003,600,000,000,000	25,000,0001,000,000	Theoretical total mass-energy of 1 gram of matter ($E = mc^2$)
150,000,000,000,00090,000,000,000,000	41,666,66725,000,000	Total energy expenditure for construction of the Great Pyramid of Giza (2485–2475 BCE)
600,000,000,000150,000,000,000,000	166,666,66741,666,666,667	Energy released by an average hurricane in 1 second
170,000,000,000,000,000600,000,000,000,000	47,222,222,222166,666,666,667	Total energy from the sun that strikes the Earth each second
190,000,000,000,000170,000,000,000,000,000	52,777,777,77847,222,222,222	2011 Tōhoku earthquake and tsunami

(continued)

Table 3.6 (continued)

J	kWh	Nature of energy release
210,000,000,000,000,000,190,000,000,000,000	58,333,333,333,352,777,777,778	Energy released by the Tsar Bomba, the largest nuclear weapon ever tested (50 megatons)
420,000,000,000,000,000,210,000,000,000,000	116,666,666,666,758,333,333,333	Electricity consumption of Norway in 2008
470,000,000,000,000,000,420,000,000,000,000	130,555,555,556,116,666,666,667	Electricity consumption of Thailand in 2008
800,000,000,000,000,000,470,000,000,000,000	222,222,222,222,130,555,555,556	Estimated energy released by the eruption of the Krakatoa volcano in 1883
3,200,000,000,000,000,000,800,000,000,000,000	888,888,888,889,222,222,222,222	Total global lightning in one year
3,550,000,000,000,000,003,200,000,000,000,000	986,111,111,111,888,888,888,889	Electricity consumption of Japan in 2008
508,000,000,000,000,000,003,550,000,000,000,000	141,111,111,111,111,986,111,111,111	Global energy production in 2010
7,900,000,000,000,000,000,508,000,000,000,000,000	2,194,444,444,444,440,141,111,111,111	Estimated energy contained in the world's petroleum reserves as of 2010
24,000,000,000,000,000,000,0007,900,000,000,000,000,000	6,666,666,666,666,670,2194,444,444,444,440	Estimated energy contained in the world's coal reserves as of 2010
29,000,000,000,000,000,000,024,000,000,000,000,000,000	8,055,555,555,555,550,6666,666,666,666	Energy yield of the known global uranium-238 resources using fast reactor technology
5,500,000,000,000,000,000,000,000,039,000,000,000,000,000,000,000	1,527,777,777,777,777,780,00010,833,333,333,333,300	Solar radiation intercepted by the Earth in one year
380,000,000,000,000,000,000,000,0005,500,000,000,000,000,000,000	105,555,555,555,556,000,0001,527,777,777,777,780,000	Total energy output of the sun each second

Table 3.7. *Units of energy measurement. The exajoule (EJ) and the terawatt-hour (TWh) are only used to describe energy use at national or global scales*

Value	Symbol	Name	kWh	TWh
10^3 J	kJ	kilojoule	0.000277	
10^6 J	MJ	megajoule	0.277	
10^9 J	GJ	gigajoule	277	
10^{12} J	TJ	terajoule	277,777	
10^{15} J	PJ	petajoule	277,777,777	0.277
10^{18} J	EJ	exajoule	277,777,777,777	277

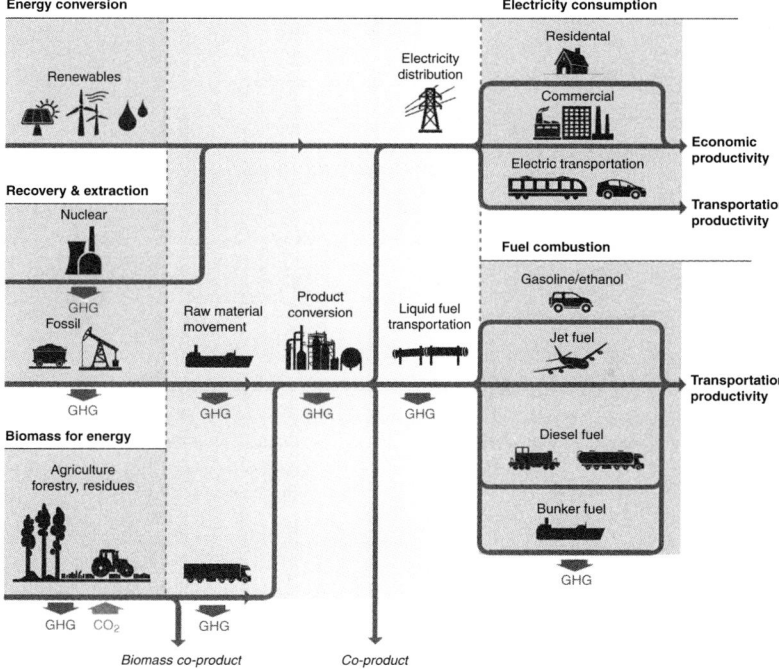

Figure 3.2. The global system for energy production and consumption. Most of the sources, processes, and methods of consumption involve greenhouse gas (GHG) emission. Biomass also involves CO_2 removal. *Source:* Arvizu et al. (2011b).

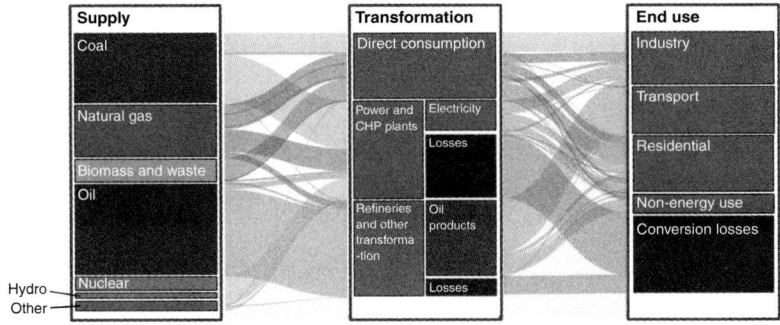

Figure 3.3. Global energy flux describing the path from energy production to consumption. The conversion and consumption pathways show that most coal goes to power plants, natural gas is equally divided between direct consumption and the residential sector, while most oil is consumed in transportation. This diagram also illustrates the global loss of energy through conversion, which accounts for more than 30 per cent of primary production.
Source: IEA website (modified).

much as one of the poorest families on Earth. Today, a comparably wealthy individual consumes 20 times as much as their poorest compatriots, and well more than 100 times as much as most rural Africans and Asians[3].

As a rule, the richer a country is, the more energy its people consume. However, culture, habits, geography, and local market forces also play a role. Top of the league table of per capita energy consumption are wealthy countries with very hot or very cold climates, such as Iceland and Qatar (see Table 3.8).

Agriculture and Food Production

Energy expenditure in food production has increased exponentially over the last 200 years. This process began with large machines that could harness the combined pulling power of several horses, was followed by the invention of motorized machines and the discovery of the oil to drive them, and finally by the development of oil-derived fertilisers and pesticides. These three factors caused an explosion in agricultural

---

[3] Data are available on the IEA (www.iea.org) and World Bank (www.worldbank.org) websites.

Table 3.8. *Energy use per capita in selected countries*

Country	MJ/capita/year
Iceland	752,912
Qatar	538,674
Kuwait	467,456
Canada	308,986
United States	295,965
Australia	224,664
Russian Federation	206,284
Germany	157,214
Japan	150,055
Israel	131,172
United Kingdom	126,106
Iran	117,147
South Africa	114,635
Spain	114,174
Venezuela	110,908
Malaysia	107,559
Argentina	77,414
China	75,655
Mexico	65,565
Brazil	56,982
Indonesia	36,174
India	24,074
Morocco	21,855
Kenya	20,013
Philippines	18,129
Bangladesh	8,583
Eritrea	5,443

Source: World Bank website (www.worldbank.org).

productivity after the Second World War, with global production increasing by 250 per cent between 1950 and 1984 (Smil 2008).

While agriculture accounts for just 2 per cent of global energy consumption (IEA website), the share invested in food production is far higher, as energy is also required to process, package, store, transport and cook food. In wealthier countries the largest expenditure

58   The Renaissance of Renewable Energy

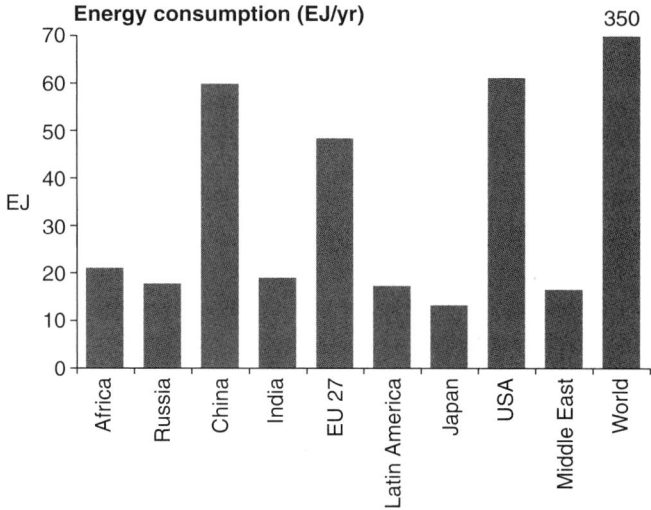

Figure 3.4. Total energy consumption in different areas of the world.
Source: IEA website.

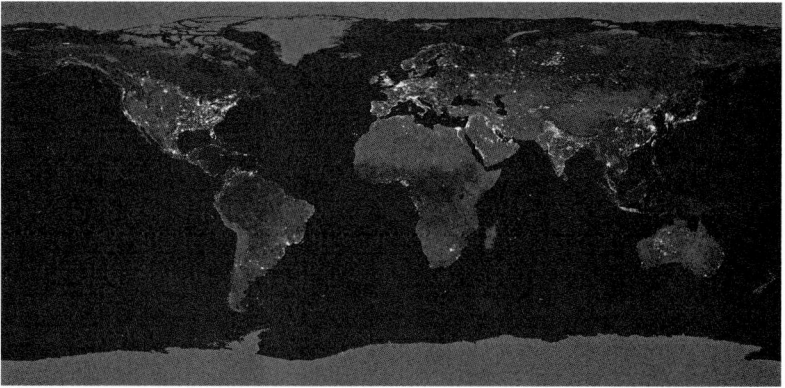

Figure 3.5. Composite satellite image showing urban lighting on Earth at night. This graphically shows where energy expenditure is concentrated.
Source: Craig Mayhew and Robert Simmon, NASA GSFC.

of energy is in processing and transport, while in poorer countries cooking accounts for the highest share of energy expenditure on food. Overall, the global food sector currently consumes 95 exajoules of energy per year, a quarter of the world's total energy consumption (FAO 2011b; IEA website).

Table 3.9. *Global energy consumption by sector*

Sector	EJ	% of total energy consumption
Industry	95.5	27.3
Transport	95.6	27.3
Residential	85.2	24.4
Commercial and public services	29.3	8.4
Agriculture/forestry	6.9	2.0
Fishing	0.3	0.1
Non-specified	5.7	1.6
Non-energy use	31.3	8.9

Source: IEA website (www.iea.org).

Home and Commercial Energy Use

In wealthier countries, access to electricity and fuels for heating, cooking and transport is largely taken for granted, and a relatively small proportion of household income is spent on energy, compared with accommodation, entertainment, education, and so on. By stark contrast, a majority of people in Africa have no access to electricity, and, like many in Asia, rely mainly on biomass (wood and crop residues) and animal dung for cooking (WHO/UNDP 2009). There are 1.3 billion people who lack access to electricity. Twice as many, 2.6 billion people, lack access to modern fuels such as natural gas and kerosene, and instead rely on traditional biomass use for cooking, causing harmful indoor air pollution (IEA 2012a; United Nations 2007).

But even amongst the wealthier societies there are major differences in the way energy is consumed in homes. The average European household consumes about 4,500 kilowatt-hours of electricity per year, whereas a household in Japan uses around 6,000 kilowatt-hours, and its American equivalent slightly more than 11,000 kilowatt-hours. The reasons for this disparity include different housing patterns, consumption habits, and attitudes towards energy use. The majority of the energy consumed in homes in wealthy countries is used to heat or cool air. Lighting, water heating, refrigeration, washing, drying, cooking and electronic devices account for the rest. While the energy demands of large appliances and boilers have been reduced thanks to improved efficiency, these gains have been more than offset by the adoption of

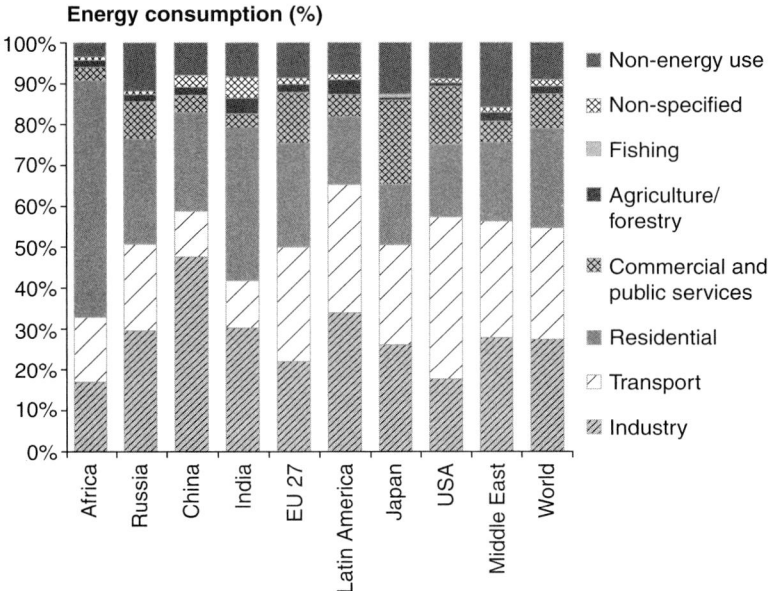

Figure 3.6. Share of energy consumption by sector in different parts of the world. Non-energy use indicates the use of fuels as raw materials, for example, oil to produce plastics or fertilizers.
Source: IEA website.

numerous new appliances, such as personal computers and mobile phones, and by the increased use of air-conditioning.

In terms of energy usage, there is not much difference between the domestic and commercial energy sectors. In both cases, energy is primarily used to heat and cool space, to provide lighting and to power electrical appliances. Not surprisingly, the bulk of commercial energy consumption occurs in wealthy societies with large commercial and service sectors. The service and commercial sectors in Japan and the United States now consume almost as much energy as their respective industrial sectors. This also reflects a global shift in industrial production to China and India, the world's emerging manufacturing giants, where ten times more energy is used in industry than in commerce (see Figure 3.6).

Industrial Energy Use

We are used to thinking of cars as machines that consume energy, but it is worth considering the amount of energy that is already 'embodied' in a car when it rolls off the production line. About 80 per cent

Table 3.10. *The amount of energy required to manufacture industrial materials from a raw natural source*

Material	Source	Embodied energy MJ/kg
Aluminium	Bauxite	190–230
Aluminium	Recycled metal	10–40
Bricks	Fired clay	2–5
Concrete	Raw materials	6–12
Copper	Ore	60–150
Cotton	Standing plant	140
Glass	Silicon sands	15–30
Iron	Ore	20–25
Lead	Ore	30–50
Oxygen	Air	6–14
Paper	Wood pulp	25–35
Polyethylene	Crude oil	75–115
Silicon (pure)	Silicon sands	1,400–4,100
Steel alloy	Iron ore	30–60
Timber	Forestry	1–3
Water	Streams	< 0.01

Source: Smil (2008).

of a car consists of metal – steel, aluminium and copper. The rest is plastics, glass and rubber. Starting with the mining of iron ore, and continuing through to the labour required for its manufacture, transport and sale, it takes about 100,000 megajoules of energy just to put an average car into a showroom. This is equivalent to roughly two years of driving. In other words, the process of making a car requires about one-fifth of the energy the car will consume in its lifetime. With computers the embodied energy is even greater, compared with the energy requirements of the device, as it takes four times more energy to produce a personal computer than to operate it throughout its lifetime (Williams 2004).

Most industrial processes require huge investments of heat and mechanical power, and almost every manufactured product undergoes several stages of energy input, starting with the raw materials (see Table 3.10). It takes about 200 megajoules of energy to manufacture one kilogram of aluminium from bauxite ore. Paper, a product so cheap and ubiquitous in wealthy countries that it is often given away for free, is also an energy glutton. Just imagine the energy required to chip, grind and boil the wood into pulp, and then to roll and dry the pulp into paper. No industrial product better illustrates the profligate use of energy in modern industrialised societies.

The energy sector is, itself, another big spender of energy. Between 10 per cent and 20 per cent of the energy contained in crude oil is expended in its extraction, transportation and refinement into lighter combustion fuels such as petrol and diesel. Though coal and gas do not need to be refined, 5 per cent to 10 per cent of their inherent energy is expended in extraction and transportation to power plants and homes. Another 10–20 per cent is used to build and run the power plants and to carry the electricity to the site of consumption. Nuclear fuel enrichment is also highly energy intensive, and nuclear power stations are so expensive to build that no company can afford to do so without government support. Some renewable technologies also have a high energy embodiment. Because silicon (the most important material in solar PV panels) requires so much energy to manufacture (about fifty times more per kilo than steel), a solar PV plant needs roughly four years to reproduce the energy expended in its construction (Desideri et al. 2012; Nawaz and Tiwari 2006), while wind turbines amortize their energy investment in about one year (Crawford 2009).

Transportation

The price of petrol and diesel is one of the most reliable barometers of fluctuations in the global energy market. This is because the transportation sector, unlike any other, is almost entirely dependent on a single energy source: petroleum. The exceptions to the dominance of petroleum in transportation are at the two extremes of technological development: the animals used to draw heavy loads in many poorer countries and the high-speed electrified railways installed in many industrialised countries (Rodrigue et al. 2013).

The transport sector accounts for roughly 27 per cent of total world energy consumption. This is likely to increase in the future as populous societies in Asia, Africa, and South America become more prosperous and mobile. Because transportation, by definition, requires a portable energy source, the opportunities for switching to renewables are fewer. The most viable alternatives to fossil fuels are biofuels, hydrogen, and electricity. Biofuels and electricity are already used extensively in transportation (biodiesel and ethanol are often mixed with petroleum, whereas electricity is used to power trains), while hydrogen is used in experimental buses and cars. Perhaps the most viable approach to reducing dependence on petroleum in the short term is the hybrid vehicle, which uses an internal combustion engine supplemented by an electric motor and battery. This combines the efficiency of electricity with the long driving range of an internal combustion engine.

3.3 Energy Reserves

Fossil Fuel Reserves

Despite her occasional calls home to my parents, I liked my elementary school teacher Miss Bissi. She seemed different from most other adults in our town; she wasn't married, she rode a bike though she could afford a car, and she told us that oil would run out in forty years. That was the time of the first oil crisis. I can remember my parents discussing the fuel shortage and whether we could take the car out on Sundays. Their crisis was our opportunity, as my friends and I turned the deserted streets of Ferrara into our own roller-skating park.

Miss Bissi (or rather her source) was wrong about oil running out in forty years, but she was right to point out that it was a limited resource. This knowledge has since become commonplace. Where doubt arises and experts disagree is on the amount of oil that remains underground. Forecasts of oil reserves have been revised several times since the 1970s, as new fields have been discovered and technology improved, allowing further extraction from existing fields. Annual global oil production actually increased from 2.9 billion tons in 1973 to more than 4 billion tons in 2010 (IEA 2012a). At current rates of production, known oil reserves will last at least fifty years (BP 2011, 2012) or perhaps as long as a century (Maugeri 2009).

Energy experts rarely talk about a moment when oil will run out, like a well that suddenly runs dry. Rather, most discussions about the future of oil focus on when production will reach its peak. From an economic point of view, this moment is far more significant than the day when the last drop is extracted. As soon as it becomes apparent that oil production is decreasing, the price of oil will soar, as speculators and governments buy up stocks. The idea of 'peak oil' was first introduced by the geophysicist Marion King Hubbert in 1956. Since then, the Hubbert Peak Curve, a parabola tracking a steep rise in oil production, its peak, and subsequent sharp decline, has become an iconic symbol of our energy predicament.

Some analysts claim that we are very close to peak oil, or even that we have reached it already. Others point out that oil production has steadily increased in recent years, so a global peak or plateau is not yet in sight (Maugeri 2011; Murray and King 2012; Zecca and Zulberti 2006). The moment of peak oil is very difficult to predict, and we will probably not know it until it has passed. There are three main reasons for this uncertainty.

First, new technology is emerging that allows far more oil to be extracted from existing fields. At present, only 10–15 per cent of the oil in a reservoir comes out spontaneously after perforation. This is known as primary production. Once the natural internal pressure is reduced, water or gas is injected into the reservoir to push more oil towards the wells. This method allows 20–40 per cent of the oil to be recovered (secondary production). More recent methods, such as the injection of chemicals or bacteria that dilute the remaining oil, have raised production to 60 per cent of the reservoir (tertiary production).

Second, many scientists and engineers are confident that new reservoirs will be discovered. However, experience suggests that we are unlikely to stumble upon any giant oil field to rival the Ghawar field in Saudi Arabia, as the rate of oil field discovery has declined over the last thirty years, despite huge advances in the technology of exploration (Zecca and Zulberti 2006). Rather, we are likely to rely more on 'unconventional' oil in Canada, Venezuela and Russia, which will become more profitable to extract as oil prices rise.

The third unpredictable factor affecting oil production is, of course, demand. Demand in the most industrialised countries has probably already peaked (BP 2011), but this has been more than offset by growing demand from rapidly growing economies such as China, India and Brazil. Demand for oil is far from equally distributed throughout the world. The wealthier industrialised countries are still the biggest consumers of oil, but several 'developing' countries are hot on their heels.

The distribution of the world's oil reserves is even more uneven. Fifty per cent of the world's known reserves are located in just five countries clustered around the Persian Gulf. Another 25 per cent are located in five other countries (see Figure 3.7). This factor, as much as impending oil scarcity, has fed the desire in many countries to reduce dependence on oil and to find alternative sources of energy.

Total world coal reserves, comprising all grades from lignite to anthracite, is currently estimated at 860 billion tonnes (BP 2011). This makes coal by far the most abundant of the fossil fuels. If we continue

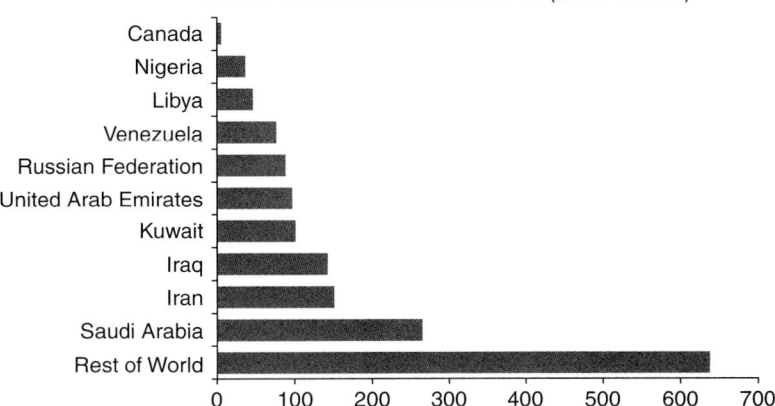

Figure 3.7. Proven reserves of oil. Canadian and Venezuelan tar sands are not taken into account (for comparison, see Table 3.11).
Source: BP (2012).

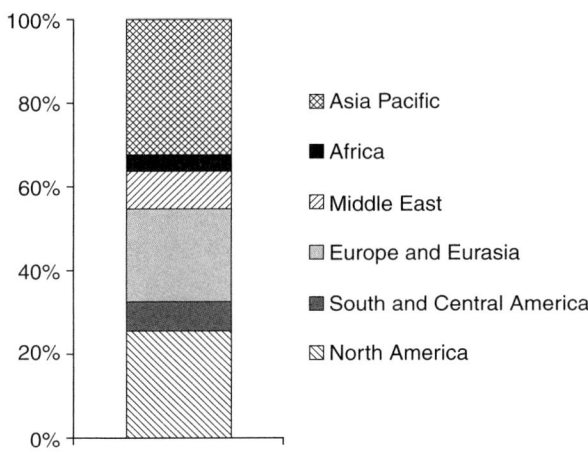

Figure 3.8. Oil consumption by region.
Source: BP (2012).

using coal at current rates, our known reserves will not run out for more than 100 years (BP 2011).

The world's reserves of natural gas are estimated at 185 trillion cubic metres, enough to meet current production levels for sixty years. Like oil, natural gas is concentrated in a small number of countries, mainly in Russia and the Middle East (BP 2011, 2012; IEA 2012a). This

Table 3.11. *Proven global reserves of fossil fuels and uranium*

Coal (billion tons)		% Global	Oil (billion barrels)	Conventional	Total	% Global
United States	237	28	Saudi Arabia	265	265	16
Russian Federation	157	18	Venezuela	77	297	18
China	115	13	Canada	6	175	11
Australia	76	9	Iran	151	151	9
India	61	7	Iraq	143	143	9
Germany	41	5	Kuwait	102	102	6
Ukraine	34	4	United Arab Emirates	98	98	6
Kazakhstan	34	4	Russian Federation	88	88	5
South Africa	30	4	Libya	47	47	3
Colombia	7	1	Nigeria	37	37	2
Rest of world	70	8	Rest of world	638	249	15
Global total	**861**	**100**	**Global total**	**1,263**	**1,653**	**100**
Natural gas (trillion m³)	**Total**	**% Global**	**Uranium (thousand tons)**	**Total**		**% Global**
Russian Federation	44.6	21.4	Australia	1,243		23.6
Iran	33.1	15.9	Kazakhstan	817		15.5
Qatar	25.0	12.0	Russia	546		10.4
Turkmenistan	24.3	11.7	South Africa	435		8.3
United States	8.5	4.1	Canada	423		8.0
Saudi Arabia	8.2	3.9	USA	342		6.5
United Arab Emirates	6.1	2.9	Brazil	278		5.3
Venezuela	5.5	2.6	Namibia	275		5.2
Nigeria	5.1	2.4	Niger	274		5.2
Algeria	4.5	2.2	Ukraine	200		3.8
Rest of world	43.5	20.9	Rest of world	426		8.1
Global total	**208.4**	**100.0**	**Global total**	**5,259**		**100.0**

Source: BP (2012), European Energy Portal (www.energy.eu).

means that, despite the advantages of natural gas over oil as a fuel source, many of the same issues of energy security apply.

Uranium Reserves

Though its use in glass making dates back to ancient Rome, uranium was not identified as an element until 1789, a few years after the discovery of the planet Uranus, after which it was named. Uranium is quite plentiful on Earth, about as common as tin or zinc, and is found, at least in small quantities, in most rocks and all seawater. At present, uranium is mined in the form of an ore. Based on current nuclear power capacity, present reserves would last another 100 years (OECD and IAEA 2010). Because of the nature of nuclear fuels, it is difficult to estimate the extent of the remaining reserves. As energy prices rise and nuclear technology improves it may, in fact, become feasible to harness a variety of other uranium sources, including seawater. This would open up the prospect of thousands of years' supply of nuclear fuel (Hansen 2008; Lake et al. 2009).

Renewable 'Reserves'

Unless we count the sun's inevitable self-destruction in 5 billion years' time as a limit, renewables are, by definition, the one energy sector with limitless reserves. Unfortunately, our success in harnessing the energy of the sun, the wind, the oceans, and the earth has been rather modest so far. Although the renewable sector has grown impressively in recent years, it is still tiny compared with the fossil fuel and nuclear sectors, and miniscule compared with the potential energy harvest from renewable sources.

A search for data on the potential of renewable energy yields a miscellany of different estimates. As with fossil fuels, many variables come into play. First, it is difficult to estimate the actual power of the energy source: how much wind blows through the atmosphere; how many waves move through the oceans; how much photosynthetic biomass grows on our planet? Second, we obviously cannot divert all agriculture to biofuel production, or cover every inch of our planet with solar panels, or install wind turbines on every mountain. Therefore, the extent to which we expand sustainable ways of producing energy will depend on our willingness to compromise over land use and, indeed, on the perceived incentive to do so. Third, the energy market will, to a large extent, determine the future of renewables;

as conventional energy sources become more expensive, renewable energy will become more attractive. Finally, a lot will depend on investment in new technology and on the determination of governments to encourage and support such investment.

Politicians are apt to wax lyrical about the potential of renewable energy. Yet a glance at the current figures for new renewables (that is, wind, solar, geothermal, and ocean power) provides a sobering counterpoint, as they account for less than 1 per cent of the global energy mix. This is the great paradox of renewable energy: practically everyone, from an oil company executive to a Greenpeace volunteer, will agree that renewables are the energy source of the future, yet no one seems to know when that future will begin.

4
How Energy Is Produced

4.1 Turning Heat into Power: The Fundamentals of Energy Generation

Apart from gravity, heat is the form of energy with which we are most familiar. As warm-blooded animals, we convert a significant part of the chemical energy in our food into heat. About half a million years ago, we also learned to generate heat from external sources by burning biomass. Since then, we have greatly expanded the range of fuels we burn to produce heat, but we still rely in most cases on combustion (the exception in our modern energy system is nuclear power, where heat is generated by exploiting a chain reaction of uranium atoms). However, while we still rely largely on combustion to generate heat, we have learned to turn heat into a far more portable and flexible form of energy – electricity.

Most ancient civilizations developed along waterways: the Mesopotamian along the Tigris and Euphrates, the Egyptian along the Nile, the Chinese along the Yellow River, the Indian along the Indus, the Roman along the Tiber. Water is essential to human life, for agriculture, fishing, and as a transportation medium. It is also valuable for washing (clothes, food, and bodies) and for industry. In the last 150 years, another inducement has been added to the list: water is an essential resource for generating electricity.

Even the most technologically advanced of modern power plants is basically a giant boiling pot. The heat generated by burning fuels or splitting uranium atoms is used to boil water, which drives steam turbines. Water is needed, not only as a 'raw material' for steam but also as a cooling agent to prevent the plant overheating. In the United States, thermoelectric plants use as much water as farms do, and fifty

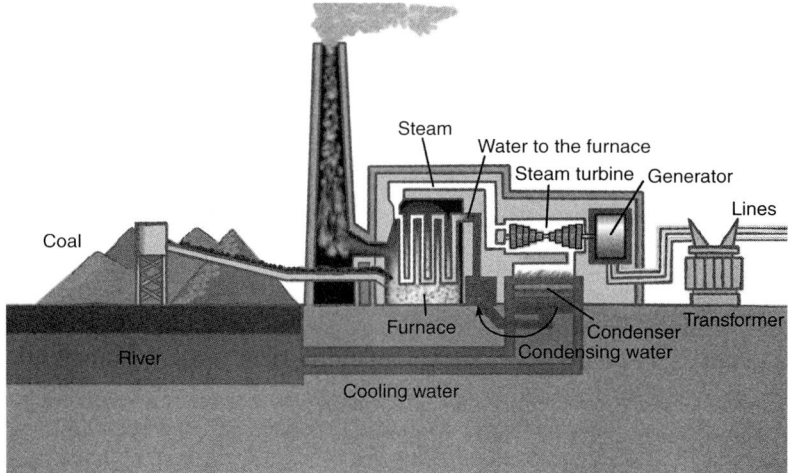

Figure 4.1. Diagram of a thermoelectric power plant, where the energy stored in the chemical bonds of the fuel (in this case coal) is converted, first into heat and then into electricity. The transformer adapts the electrical current to that of the grid.
Source: Tennessee Valley Authority (modified).

times as much as homes do.[1] That is why power plants tend to be built next to a natural source of water.

Although the technology used in power plants is highly complex, the process of energy conversion from a fuel to an electrical appliance is surprisingly simple. Whether fuelled by coal, oil, natural gas or uranium, a power plant converts the energy stored in the fuel into heat. The heat released is immediately used to turn water into steam at temperatures of around 500 degrees Celsius. The superheated steam is carried, via a network of pipes, to a turbine, where it drives blades that are connected via a shaft to a generator. Technologically speaking, this generator is like a supersize bicycle dynamo, in which rotating magnets convert the mechanical energy of the rotating shaft into electricity. The electricity enters the grid and travels at the speed of light to homes, offices and factories, where it is again converted into light and heat or used to power machines and electronic devices.

Some renewable technologies (biomass and concentrated solar power) exploit the same 'boiling pot and dynamo' principle, converting heat to motion and motion to electricity, though the scale is usually

[1] See www.ucsusa.org

much smaller. While a coal power plant may exceed 1,000 megawatts of capacity, biomass power plants with more than 100 megawatts of capacity are rare, as are concentrated solar power (CSP) plants with more than 5 megawatts. Wind and hydroelectric plants dispense with heat generation entirely, though the latter stage (the conversion of mechanical to electrical energy) is similar.

Energy, Power and Capacity

Which contains more energy – a plate of cookies or a small hand grenade? The answer, counter-intuitively, is the plate of cookies. One hundred grams of cookies have an energy content of about 1,500 kilocalories (6,279 kilojoules), three times more than the same amount of TNT. The TNT releases a smaller quantity of energy than the cookies, but does so in a far shorter period of time. It contains less 'energy' than cookies but has much more 'power'.

The joule (J) is the unit used to measure energy, while the watt (W), defined as 1 joule per second, is used to measure power (Chapter 1). However, in addition to the joule, there is alternative unit measure of energy, which will be familiar to anyone who has had the privilege of receiving an electricity bill: the kilowatt-hour (kWh).

Lightning has too much power

I thought it had too much energy

Whatever it has, it's too much!

Every externally powered device, from a domestic electric heater to a nuclear power plant, has a nominal power capacity, describing the maximum power at which the device can operate continuously. A washing machine with a nominal power intake of 1,000 watts changes its energy consumption during operation. It goes from 100 watts when pumping water to maximum power during spin-dry mode. For technical and economic reasons, power plants do not always operate at full capacity. At night, output is usually reduced because demand is lower. Moreover, every facility needs to be switched off for routine maintenance, or is liable to experience machine failures. Therefore, measured over the course of a year, a power plant will have an output well below its capacity.

A highly efficient nuclear power plant operates at about 90 per cent of its capacity. This percentage is referred as the 'capacity factor'. Nuclear power plants have the highest capacity factor of all electricity generation facilities, while renewables, such as solar and wind power, have the lowest. Of course, the capacity factor also depends on several other factors, including the quality of the fuel, the reliability of the source, the technology, and whether the plant was intended for base- or peak-load supply (see Table 4.1).

Base Load and Peak Load

Over the course of twenty-four hours, our homes and our bodies experience ebbs and flows of energy consumption. In the evenings, homes are a hive of activity, as parents fuss to make dinner, children squirm while trying to finish their homework, the washing machine does its thing, and grandma knocks out an email on an arthritic keyboard. During the night, the household's energy expenditure dips low, represented by the nasal purr of the fridge and some throaty snoring from the master bedroom. In the morning we wake up a bit later than the central heating does, and new energy demands kick in.

Electricity utilities cope with these variations in two ways. First, they create what are called load profiles. These are graphs showing variations in demand for electricity (or electrical load) within a particular network or by particular consumer groups over time (see Figure 4.2). Luckily, electricity demand patterns are quite predictable, so power utilities can plan how much electricity to generate at any given time. As we have seen, electricity is very difficult to store in large quantities, so excess electricity is likely to go to waste. Utilities also take advantage of the law of supply and demand, and charge different rates depending on the level of demand.

The second way utilities adapt to fluctuating demand is by generating power in different types of facilities: base-load and peaking plants. Base-load plants are those facilities that meet a region's continuous power demand, and they generate power at a relatively constant rate. Nuclear and coal-fired plants are most commonly used

Table 4.1. *Capacity factor ranges of the main technologies for electricity production*

Source	Technology	Capacity factor (%)
Nuclear	Nuclear power plant	86–95
Coal	Coal-fired power plant	65–70
Hydropower	Dam and runoff	30–60
Wind energy	Onshore	20–40
	Offshore	35–45
PV	Residential rooftop	10–20
	Large-scale PV	15–27
CSP	CSP plant	25–75
Solar thermal	Domestic and large-scale	4–13
Bioenergy	Solid biomass	70–80
	Domestic pellet heating	13–29
	Municipal solid waste	80–91
	Anaerobic digestion	68–91
Geothermal	Electricity	60–90
	Heat	25–30
Ocean energy	Tidal	22.5–28.5
	Wave	25–40

Note: Plants that operate around the clock, such as nuclear or fossil fuel plants, have generating capacities much higher than those that can only supply electricity when the sun shines or the wind blows.

Sources: McGowin (2008), Bain (2011), Bruckner et al. (2011), Arvizu et al. (2011a), Kumar et al. (2011), ETSAP (2010), Lewis et al. (2011), Smil (2012).

to satisfy base demand. Among renewables, geothermal, biogas, and biomass plants are also suitable for this purpose. Since they are designed for maximum efficiency and operate almost continuously at high output, base-load plants have the lowest costs per unit of electricity.

Peaks (or spikes) in demand are handled by smaller, more responsive plants. These are typically natural gas and hydroelectric plants that can be switched on and off very quickly in case of sudden unplanned demand, such as during heat waves (when there is a spike in air-conditioning demand) or when base-load plants break down. Peaking plants may operate for a few hours a day, or even just a few times per year, and the power they generate commands a higher price per kilowatt hour than base-load power.

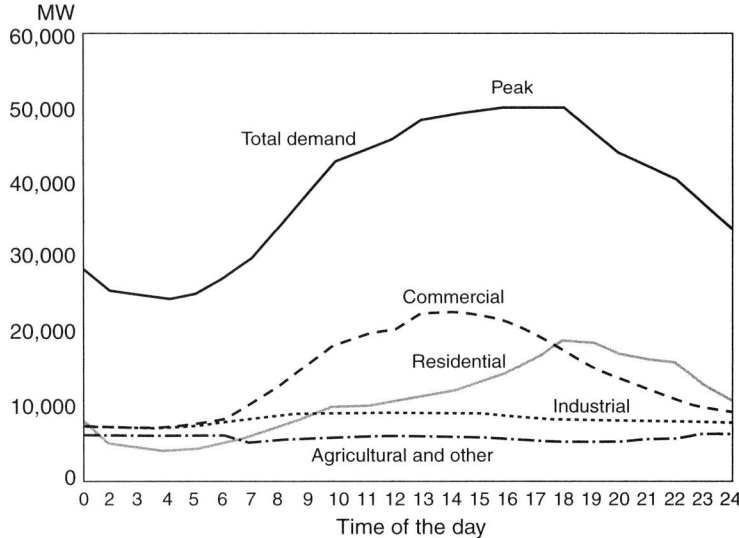

Figure 4.2. Typical daily load profile on a hot summer's day in California. Industry and agriculture tend to use electricity in a more uniform way; commercial use tends to peak during central hours, while residential demand peaks in late afternoon when most appliances are in operation. While the aggregate demand evens out the individual peaks somewhat, the daytime load is still twice the night-time load.
Source: Lawrence Berkeley National Laboratory.

Stages of Technology Development

As we have seen, conventional power plants come in a variety of shapes and sizes, and fulfil different needs within the energy market. However, the core principle upon which they operate has not radically changed since James Watt's day, as heat is generated to boil water, and the resulting steam is converted into mechanical energy. Of course, we have discovered many new fuels in the last two centuries (including an entirely new type of fuel, uranium), and in the last hundred years we have added a further link to the conversion chain: electricity. However, these days, the really exciting developments are taking place in the renewables sector. It is here that the imaginations of modern-day Edisons have room to roam.

Every technology has its own evolutionary path. It starts in the mind of a scientist and, if proven technically and economically viable, it may one day reach the market. Between these two stages lies a spectrum, which we might crudely divide into four stages:

development, demonstration, deployment, and diffusion (IEA 2008).

At the first stage (research and development), the technology has shown initial promise in a laboratory, but technical and cost barriers remain. It is still too early to judge whether it will be commercially viable. At the second stage (demonstration), the technology has been shown to work. However, costs are usually still high, and external (including government) funding is usually needed to support the demonstration. At the third – deployment – stage, the technology operates successfully but may still be in need of financial supports or subsidies. With increasing deployment, costs progressively decrease and the technology may eventually reach the final stage (diffusion and commercialisation) when it is competitive in some or all markets, either on its own terms or supported by government subsidies.

> **The Potential of a Technology**
>
> When we consider the future potential of different energy sources, it is important to be aware that there are different types of potential. In the case of fossil and nuclear fuels, their potential is directly linked to the amount of oil, coal, gas, or uranium ore that can be extracted from the Earth's crust. With renewables, the situation is more complicated, and we need to distinguish between the following types of potential.
>
> **Theoretical potential** provides the most optimistic forecast. It only takes into account natural and climatic restrictions, such as the amount of wind or sunshine available at different locations. At the poles, solar panels would not work at all in winter, and in deserts very little water is available for hydroelectric or biomass projects.
>
> **Technical potential** is the theoretical potential minus geographical and technical limitations such as conversion efficiencies. It is not practical to install solar panels in the middle of the ocean, while wind turbines on the top of a mountain would be difficult to connect to the grid.
>
> **Economic potential** is the technical potential minus possible limitations attributable to cost factors. For example, it is more economical to install and maintain wind turbines on land than at sea, even though wind strengths at sea are usually greater.

(continued)

> **Market potential** reflects the total amount of energy from a particular source that can be sold in the market, taking into account demand, competing technologies, costs and subsidies, and any market barriers. Usually the market potential is lower than the economic potential. Yet, since markets provide opportunities as well as barriers, the market potential may, in theory, be larger than the economic potential, as the recent boom in renewables shows.

4.2 Hydropower

Water is fluid, soft, and yielding. But water will wear away rock, which is rigid and cannot yield.

– Lao Tzu

Though presumably unaware of the chemical properties of its molecules, the Chinese philosopher Lao Tzu (sixth century BCE) understood that water is highly cohesive yet loosely bound; dense yet fluid. Because water is 800 times denser than air, it is a far more potent conduit of energy. The most destructive of natural disasters involve water. Even where the original cause is not hydrological – as in the case of hurricanes or earthquakes – it is usually the resulting floods or tsunamis that wreak the greatest damage.

The earliest uses of the force of flowing water were to irrigate croplands (waterwheels raised water to higher ground) and to drive machines, mainly grain mills.[2] Despite numerous improvements over the centuries, there was little substantial change in the way waterpower was harnessed until the nineteenth century, when water turbines were invented. A water turbine essentially replaces the millstone in a watermill with a generator. Water turbines emerged at roughly the same time as steam turbines, and indeed many of the world's first power stations were driven by water.

From modest beginnings in 1878, when the English industrialist William Armstrong harnessed flowing water to power a single lamp in his home, hydropower developed rapidly. Just four years after Armstrong's breakthrough, Thomas Edison, that remarkable pioneer of all things electrical, opened a 25-kilowatt hydroelectric power plant in Appleton, Wisconsin. Over the next two decades, the technology

---

[2] Waterwheels for irrigation and milling were used in the Mediterranean as long ago as the third century BCE.

spread rapidly, and throughout the twentieth century hydropower plants led the way in terms of scale. The Hoover and Grand Coulee dams, completed in 1936 and 1942 with capacities of 1.4 and 6.8 gigawatts, respectively, dwarfed contemporary power plants. Even today, the world's largest hydropower plant, China's Three Gorges Dam (see Figure 4.5), has a capacity several times greater than the largest fossil fuel or nuclear power plants.

Early hydroelectric plants were built in mountainous regions to take advantage of fast-flowing streams and waterfalls. Later, dams were built to create a reservoir of potential energy that could be tapped on demand. Such projects not only allowed for electricity generation on a grand scale but also involved enormous investment and environmental disruption. Many countries, particularly those with few native fossil fuel reserves, came to rely heavily on hydropower to satisfy their electricity needs, and, in many cases, hydropower schemes kickstarted the process of electrification.[3] To this day, hydroelectric plants have iconic status in many developing countries, representing both the blessings and the blight of industrial development.

Power Generation in All Shapes and Sizes

The technology of hydropower is relatively simple. The kinetic energy of flowing water drives a turbine, thereby generating electricity. Like wind power, hydropower is an indirect form of solar energy. Without the constant replenishment of the river source through evaporation and rainfall, the river and the reservoir would quickly dry up.

There are two main types of hydropower plant: dams and run-of-the-river schemes. Dammed hydropower relies on a reservoir that builds up behind the dam, while run-of-the-river schemes, which are small-scale by nature, take advantage of the natural flow of the river and a very small reservoir, known as pondage (see Figure 4.4). The capacity of a dammed hydropower plant depends mostly on the volume of water in the reservoir and on the size of the dam. Unlike other renewable energy sources, hydropower lends itself to megaprojects. The world's four largest power plants are all hydropower facilities, with capacities from 22 to 8 gigawatts. However, such mammoth hydroelectric projects are limited by the number of suitable rivers,

---

[3] In the United States, by 1920 40 per cent of all electric power was hydroelectric. In Ireland, a single 85-megawatt hydroelectric plant, completed in 1929, satisfied, for its first few years of operation, the fledgling state's entire electricity demand.

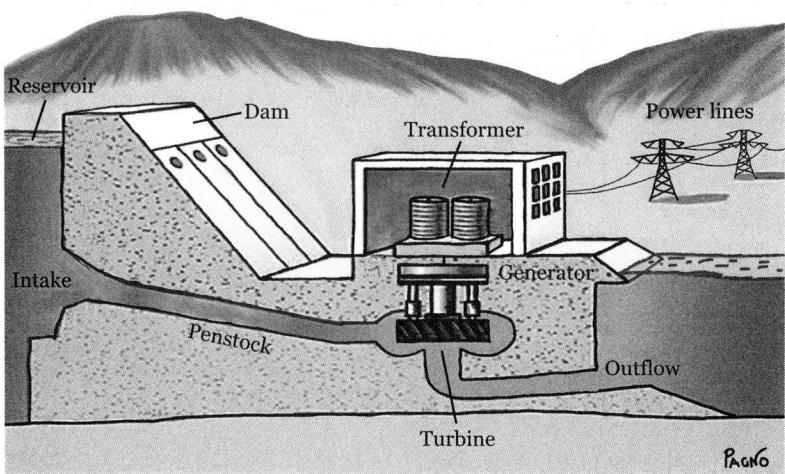

Figure 4.3. A dammed hydropower plant. Large pipes, known as penstocks, channel water from the reservoir, through the dam, to the turbine.

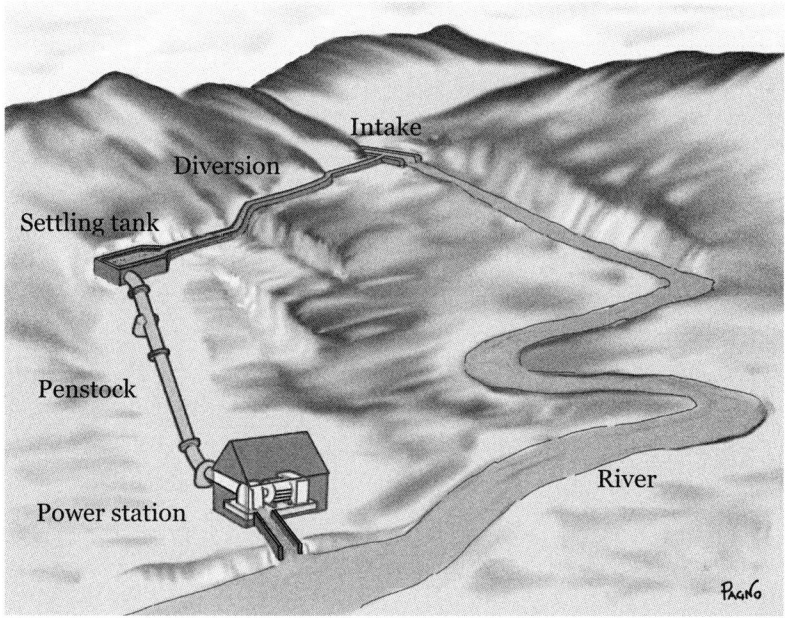

Figure 4.4. A run-of-the-river hydropower scheme. An artificial diversion brings part of the river's water to a settling basin. A penstock connects to the power station where the turbine-generator system produces electricity. The water is then returned to the river.

Figure 4.5. The Three Gorges Dam hydroelectric plant on the Yangtze River in China. With a capacity of 22,500 megawatts, it is by far the largest power plant in the world.
Source: Le Grand Portage at Wikimedia Commons.

not to mention by the social and environmental disruption they cause (Chapter 6).

Despite the conspicuous existence of giants like the Three Gorges Dam, the vast majority of hydroelectric plants are small in scale (within the European Union alone there are more than 20,000 small hydropower plants)[4]. Small hydro tends to serve a rural community or a single industrial plant. The major advantages of such plants is that they can be built in isolated areas that are not connected to the grid and have environmental and social impacts that are negligible compared with large-scale schemes.

In Rangkhani, a remote mountain region of western Nepal, life revolves, as it has done for centuries, around subsistence farming and cottage industry. In 2001, a 26-kilowatt micro-hydro project transformed daily life in Rangkhani by providing electricity to more than 1,000 people, their their homes and small enterprises. Local incomes and educational opportunities have increased, while health impacts from dirty kerosene lanterns have fallen (Pottinger 2012; Yee 2012). Such schemes, with a capacity of less than 100 kilowatts, are referred to as micro-hydro plants.

---

[4] See http://www.erec.org/renewable-energy/hydropower.html

Figure 4.6. Tungu-Kabri micro-hydro power project in Kenya. At bottom left is the diversion that brings water to the small power station.
Small projects may provide poor communities in rural areas with an affordable, easy to maintain solution to their energy needs.
Source: www.practicalaction.com.

They can provide power to an isolated home or small community and are usually not connected to a power grid. In countries such as Nepal and Nigeria that have a low rate of electrification and an inefficient grid, but large and untapped hydropower resources, small hydro projects are more appropriate than large dams for meeting energy demands. The United Nations estimates that each new micro-hydro system supports forty new businesses (UNDP 2012).

Micro-hydro systems are also frequently used in conjunction with solar systems, because in tropical and subtropical regions, dark rainy seasons (good for hydropower) are followed by hot dry ones (good for solar).

Pumped Storage

In addition to its use in generating electricity, hydro also plays an important role in energy storage. This involves humans assuming the role normally played by the sun, as water is pumped uphill and held in a reservoir.

Though this process entails a net loss of energy, at present it is the most effective way of storing energy on a large scale. It makes sense economically for two main reasons. First, most conventional power plants operate at the same capacity around the clock (this is especially true of nuclear power), even though demand for electricity varies greatly between day and night. Pumped storage therefore provides a way to take advantage of the surplus power produced at night. This energy can then be released and sold at a higher price during the day. Second, pumped storage works well alongside other renewable technologies. Wind and solar power are, by their nature, highly intermittent; solar power only works during the day, while wind is quite unpredictable. By combining wind and solar plants with pumped water storage, it is possible to ensure a steady flow of electricity into the grid.

Benefits and Costs of Hydropower

Hydropower is a predictable energy source that serves the two extremes of the power market: large-scale production to meet peak demand and small-scale production to serve isolated communities. At the crossroads of two major issues for sustainable development – energy and water – hydroelectric reservoirs often provide other services in addition to electricity supply, such as a secure water supply for drinking, irrigation and flood control. Hydropower is also technically mature, and it has one of the best conversion efficiencies (about 90 per cent) of all energy sources. While they require high initial investment, hydroelectric plants have a long life span with relatively low operation and maintenance costs.

Many parameters, including the high costs of structures such as dams and canals, affect the cost of hydropower. However, the market price of electrical power is not the sole determinant of the economic value of hydropower projects. Hydropower plants designed to meet peak electricity demands may have relatively high costs, but the cost per unit of power during periods of peak demand is also higher. Moreover, hydropower projects may provide multiple services in addition to the supply of electric power.

The cost per kilowatt-hour of hydroelectricity is 5–40 U.S. cents for off-grid micro applications (< 1 megawatt), 5–12 U.S. cents from small hydro (< 10 megawatts), and 3–5 U.S. cents (2005) per kilowatt-hour for large hydro (> 10 megawatts). In the case of large projects, the technology is therefore already competitive with fossil fuels (Kumar et al. 2011; REN21 2012).

The Potential of Hydropower

While large-scale hydropower was pioneered in Europe and North America, with groundbreaking projects such as the Hoover Dam in Nevada, the current emphasis in these regions is on small-scale projects, as the potential of the major waterways has already been tapped. Today, the biggest hydropower projects are being built in Asia and South America.

Hydropower is the only renewable energy sector with an output comparable to conventional energy sources. Indeed, it is the world's third most important electricity source after coal and natural gas. In 2010, global capacity was around 1,000 GW with an output of roughly 12 exajoules per year (3,400 TWh per year), equivalent to 2.3 per cent of primary energy production, 16 per cent of global electricity production (IEA 2012a; JRC 2011; REN21 2012; Chapter 7, this volume).

According to the *International Journal on Hydropower and Dams*, the total worldwide technical potential for hydropower is 52 exajoules per year (14,576 TWh per year) or roughly 10 per cent of current global primary energy (IJHD 2010). Estimates of undeveloped capacity range from about 47 per cent in Europe and North America to 92 per cent in Africa. Asia (led by China) and Latin America (led by Brazil) are the new leaders in large-scale hydroelectric development, but Africa has the most untapped resources.

Additional power may come from upgrading old power stations. Compared with new power plants, such projects are cheaper, quicker to develop and usually have lower environmental and social impacts. Furthermore, most of the world's 45,000 large dams are used for non-energy purposes (e.g., irrigation, flood control, navigation and urban water supply schemes), with only 25 per cent used for hydropower. Significant potential can be developed from existing infrastructure that currently lacks generating units (e.g., barrages, weirs, and irrigation dams) by adding new hydropower facilities (Kumar et al. 2011).

4.3 Wind Power

Why Dogs Like to Sleep on the Couch

Atoms are like people on a dance floor. As they dance to the music, they naturally bump against one another; to create more space to dance, they move apart and occupy a greater space. In a similar way, when a substance is heated, its atoms vibrate, move apart, and the substance expands to fill a greater space. We can observe this principle in a hot-air balloon. The air inside the balloon is heated using a burner, causing it to expand. This air becomes lighter (less dense) than the air outside, causing the balloon to rise.

At the north and south poles the air is cold, therefore dense or compressed. Hence, it is referred to as high air pressure. At the equator, the opposite applies: the air is warm, its atoms more sparse, and so we call this a low-pressure area. Like the dancers who try to nab a spot on the dance floor with a bit more wiggle room, air automatically moves from an area of high pressure to one of lower pressure. This gives rise to the phenomenon we call wind. We can see this principle in action in our own living rooms; when we open the window on a winter's day, the cool air from outside rushes into the warm room, creating a current of air or a draught. This cool air falls, which explains why some people like to have a blanket over their legs and feet, and why it is not unusual to enter the living room in the morning and find the dog on the couch.

From Draughts to Hurricanes

The greater the difference in pressure between two air masses, the faster will be the speed of the wind. A draught is perhaps the least powerful and most localized of all wind types. Another localized, but more powerful movement of air is a breeze (see Figure 4.7). Breezes typically move from land to sea during the day (when the land is cooler than the sea) and from sea to land at night (when the sea is cooler than the land). The reason for the difference in temperature between land and sea is that soil temperatures can vary greatly over the course of a day whereas seawater maintains a more constant temperature. You will never burn your feet paddling in the sea, but you may on a beach in summer.

This principle also applies to winds at a continental scale (trade winds, monsoons, polar winds, etc.). For example, during the South Asian winter, the land north of the Himalayas cools, and cold, dry air moves southwards towards the warmer Indian Ocean. This causes what is known as the 'dry monsoon', affecting northern India, Pakistan, and Bangladesh. During the summer months, as northern Asia becomes hotter than the Indian Ocean, moisture-laden air is drawn from the ocean towards the hot plains of northern Asia. This causes the 'wet monsoon', which particularly affects southern India, Sri Lanka, and

Figure 4.7. A sea breeze is caused by the movement of cool air from sea to land during the day and from land to sea at night.

Indonesia. Similarly, superheated air deep in the North American continent draws cooler air from the nearby Atlantic Ocean. This air movement is most powerful in the Gulf of Mexico, which is why the Caribbean islands, Central America, and the southeastern United States are regularly hit by hurricanes.

From Windmills to Wind Farms

Wind-propelled boats plied the Nile as early as 5000 BCE. By 200 BCE, simple windmills in China were pumping water. By the eleventh century CE, wind was used extensively in the Middle East to mill grains and spices. Around this time, merchants and crusaders brought windmill technology to Europe, where it was enthusiastically adopted. No region was more profoundly influenced by early wind power than the Netherlands, where the windmill was adapted for draining lakes and marshes in the Rhine delta.

Today's windmills, instead of pumping water or milling grain, generate electricity. That is why engineers, unlike the general public, talk about wind turbines rather than windmills. Only in developing countries, or as a tourist attraction in places such as Holland, are windmills used to perform mechanical work.

Some regions are windier than others. However, the windiest sites are not necessarily the best suited to exploiting wind energy. The Caribbean region, for example, though famous for its hurricanes, is not an ideal location for wind turbines. This is because the winds are seasonal and too powerful to exploit, at least using the technology available today. Stable and moderate winds are best suited to wind power. Most modern turbines begin generating electricity at a wind speed of 3 metres per second, but cannot operate above 25 metres per second. This is because the tower and turbine may be damaged at the higher wind speeds. Temperate zones, such as northern Europe, the cone of South America, and New Zealand are the best locations for wind energy.

Wind Turbines

As a child growing up in Ireland, I made an early acquaintance with the power of wind. I learned to cycle when I was about six, and from that time onward this became my main mode of transport. In Ireland, there is no getting away from the wind. It is as defining a feature of life as the sun in the Sahara or snow in Greenland. Whether I had the wind at my back or in my face cycling to school meant a difference between arriving in time or suffering the wrath of Mr. Kelly, the vice principal, who lay in wait for latecomers. I understood that wind, depending on its direction, may be a hindrance or a help. Those who have designed windmills – from the earliest mechanical devices to the latest multi-megawatt turbines – have also had to contend with this characteristic of wind.

The challenge for engineers was to devise a windmill whose blades will turn whatever the wind's direction. They solved this problem by exploiting two forces produced by air movement: drag and lift. Drag is the wind resistance that I experienced as a young cyclist. Lift is the 'pull effect' created when the air pressure on one side of an object is greater than on the other. There are several ways to explain how lift works. In the end, all of these explanations boil down to the fact that the flow of air around an airfoil (or wing) generates an increase in air pressure under the wing and a decrease in pressure above it, causing the wing to lift (see Figure 4.9). Thanks to this effect, birds and, by imitation, airplanes are able to fly. The faster the airflow, the stronger are

Figure 4.8. We are familiar with what happens when one solid object hits another. This effect is particularly evident in ball sports where players use one object (such as a bat) to move another (the ball). However, when fluid matter (a liquid or a gas) passes around a solid object, the reaction is more complex.
Sources: Grunkcommons at Wikimedia (left); Robert Swier at Flickr.

the forces acting on the airfoil. When the speed drops below a certain level, the lift is not sufficient to sustain the wing and the bird or plane will descend towards the ground.

The most intuitive method of harnessing wind power is with a turbine that turns like a carousel on a vertical axis. The Finnish engineer Sigurd Savonius perfected this design in the early twentieth century. Because the vanes are curved, one side catches more drag than the other does. This causes it to turn, regardless of the wind's direction. The Savonius wind turbine is used to this day, most commonly

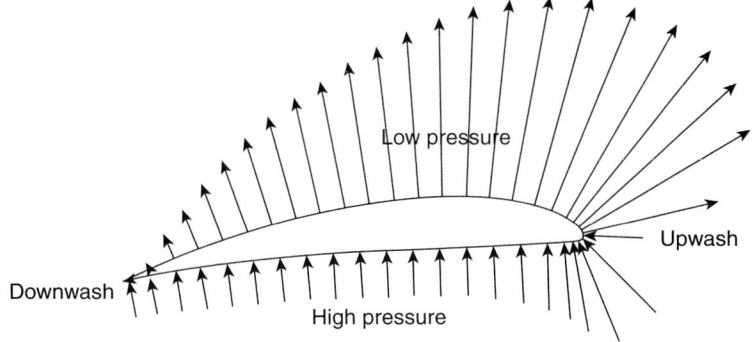

Figure 4.9. Forces that act on an airfoil, causing lift.

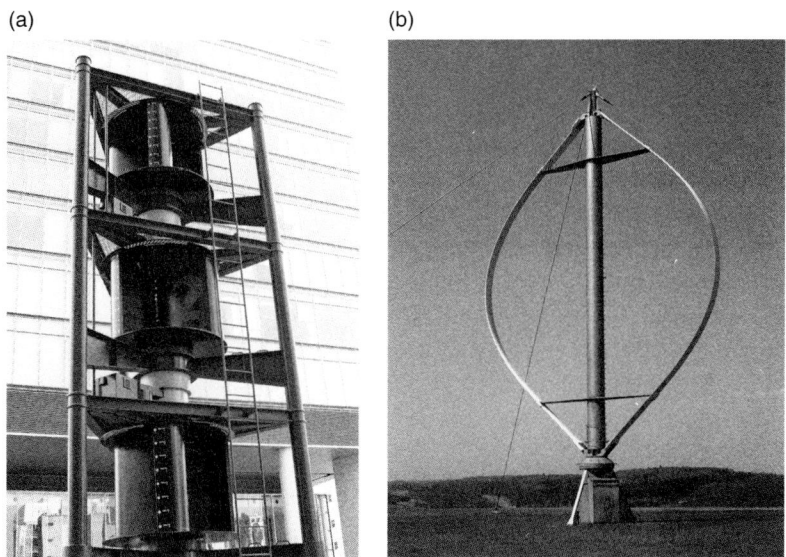

Figure 4.10. Savonius (left) and Darrieus (right) vertical-axis wind turbines.
Source: aarchiba (left) and Toshihiro Oimatsu (right) at Wikimedia Commons.

as part of a cooling device on the roofs of vans and buses. Another relatively common type of vertical-axis wind turbine (VAWT) is the Darrieus turbine, also named after its inventor. The Darrieus is more mysterious than the Savonius; indeed, without an understanding of aerodynamics it's hard to understand how it turns at all. VAWTs are mostly confined to small-scale applications, such as rooftops and small concentrated wind farms.

88 The Renaissance of Renewable Energy

Figure 4.11. Horizontal-axis turbines. The famous Don Quixote–style windmill in Spain, a small-scale turbine (1 kilowatt) for localized electricity production, and a modern wind farm consisting of several wind turbines, each with a capacity of 1 megawatt.
Sources: Jan Drewes at Wikimedia Commons (top left); Eclectic Energy Ltd. (top right); Gian Andrea Pagnoni (bottom).

While the Savonius wind turbine relies largely on drag, and the Darrieus on lift, the most commonly used type of wind turbine – the horizontal-axis wind turbine (HAWT) – employs a combination of both forces. The HAWT is like an outsize version of the pinwheel toys many of us played with as children. Commercial HAWTs vary in size from the small turbines installed on sailing boats with a rotor diameter of less than 1 metre that generate less than 1 kilowatt, to giant turbines more than 100 metres tall generating up to 7 megawatts of power.

When discussing the mechanics of wind turbines with my friends I've been amazed to discover that many of them believed that they

Figure 4.12. The three blades of a wind turbine en route through Edenfield (UK) to a wind farm.
Source: Paul Anderson at Wikimedia Commons.

function in much the same way as a waterwheel, with air substituted for water. They believed that the wind hit the rotors of the wind turbine from the side. This reminded me of my own surprise as a child when my toy windmill worked far better when I faced the wind than when I blew at it from the side. Essentially, a wind turbine works like a fan in reverse; instead of moving air, it is moved by the air.

The performance of a wind turbine depends on four factors: the size of the rotor (the combined hub and blades); the number and shape of the blades; the shape of the terrain; and the presence or absence of obstacles such as houses, trees, or hills. The turbines used in wind farms usually have three blades and always face the wind, thanks to sensors that automatically redirect the rotor. From the centre of the rotor a shaft enters the main body of the turbine (known as the nacelle) and is connected to a generator, which produces electricity in much the same way as a bicycle dynamo (see Figure 4.13).

This simple explanation conceals the fact that a wind turbine is a highly complex machine. Not only does it contain a variety of mechanical and electronic components that convert the kinetic energy of the turning rotor into electricity, the blades themselves are a marvel of engineering. They are usually made of glass fibre, a material that is both light and durable, and are contoured to take maximum

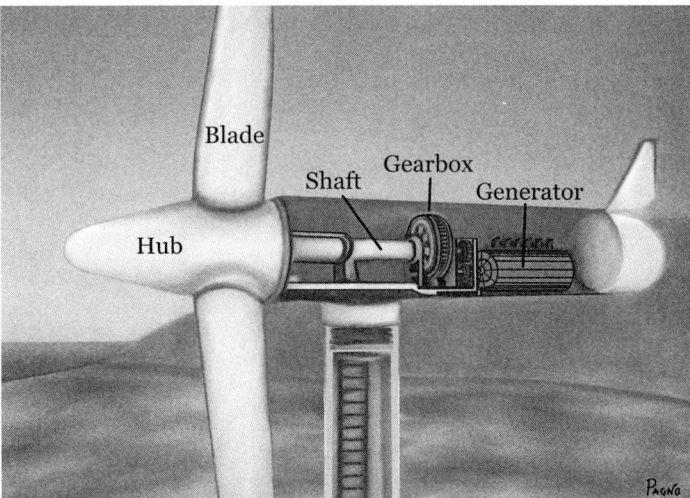

Figure 4.13. The inside of a wind turbine.

advantage of both drag and lift. Turbines are also equipped with protective features – particularly braking systems – to avoid damage at high wind speeds. The rotor is mounted on a tower high enough to avoid the air turbulence caused by obstacles.

Wind speed is measured in metres per second (m/s) or kilometres per hour (km/h). It is also measured using the Beaufort scale, ranging from 1 (a calm day) to 12 (hurricane). This scale classifies wind speed not only in terms of its actual speed but based on its effects (e.g., wave height at sea, destruction on land). The optimal conditions for wind power generation vary depending on the size and type of turbine. Small turbines, including all VAWTs and domestic HAWTs, begin to generate electricity in light winds of 2–3 metres per second (force 2 on the Beaufort scale), operate optimally at medium wind speeds of 6–10 metres per second (force 4–5 on the Beaufort scale), and cut off at winds above 25 metres per second (force 10 or higher) to prevent damage to the turbine. The largest turbines begin to generate electricity at wind speeds above 5 metres per second, operate best at speeds of 7–15 metres per second, and cut off at speeds above 25 metres per second.

Most medium-sized and large turbines feed the electricity they generate directly into the power grid. This is because wind, like solar power, is highly intermittent, and at any given time a turbine is likely to produce either too much or too little power for any single household, farm, or factory. Economy of scale is also an important factor in

Figure 4.14. A technician controls the installation of the generator in the nacelle of a wind turbine. After this, the rotor will be assembled. The holes in the hub section indicate where the blades will be attached. *Source:* Paul Anderson at Wikimedia Commons.

wind energy. A single turbine requires less space, blocks the flight path of fewer birds, and is more economical to install and maintain than numerous smaller turbines. That is why we are likely to see the trend towards larger, more powerful turbines continue. Whereas in 2005 the average output of turbines installed in commercial wind parks was 700 kilowatts, by 2010 this had increased up to 3 megawatts. Currently the world's largest wind turbine is the Enercon E-126, installed in several wind parks in Germany and Belgium. It has a rotor diameter of 126 metres, stands 198 metres tall (higher than the Great Pyramid of Giza) and has a capacity of more than 7 megawatts, capable of satisfying the power needs of about 5,000 European homes.

Current wind turbine technology has been developed largely for onshore applications. However, as installation technologies have improved, there has been a move towards 'offshore wind farms'. These are constructed in seas and lakes, usually a few miles from the shore. The advantages of offshore wind farms are steady winds, absence of obstacles, and no interference with landscape, farming or human settlement.

Small wind turbines tend not to be connected to the power grid. Indeed, these machines are often installed in places where access to the grid is impractical or expensive. They are therefore most commonly used to charge batteries for caravans, boats, or mobile health centres, and to supply electricity to remote areas (isolated settlements and farms, weather stations, etc.). Fixed small turbines are best installed in areas that receive above-average wind (higher than 6.5 metres per second annual average), and need to be accompanied by large accumulators (batteries) to compensate for inevitable periods of calm.

The Benefits and Costs of Wind Power

There are few technological or economic barriers to wind power. It is already considered commercially mature, meaning that under favourable conditions it can compete with conventional energy sources. As demand for renewable energy grows and storage technologies improve, the wind power sector is likely to expand.

However, wind energy also has several characteristics that pose challenges for electricity systems – mainly that output is quite unpredictable and turbines tend to be clustered in certain locations (Wiser et al. 2011). Some countries have already experienced wind energy surpluses at certain times. Denmark has, on occasion, paid other countries to take its excess power supply, and in west Texas surpluses of wind occasionally result in negative pricing. According to Jens Moller Birkebaek, vice president of the Danish grid operator Energinet, if significant changes to the electric system are not made, future supply might eclipse demand several times a year (Kanter 2012). New transmission infrastructure, both on- and offshore, will be required to access areas with higher-quality wind resources.

Though small-scale wind turbines (1–20 kilowatts in size) do not benefit from the economies of scale that have helped reduce the cost of larger wind turbines, they can be economically competitive in windy areas or in areas that do not have access to the grid.

Generation costs for onshore wind power plants in good to excellent wind conditions average from 5 to 10 U.S. cents (2005) per kilowatt-hour, reaching 15 U.S. cents per kilowatt-hour in lower resource areas. Offshore wind energy has typical generation costs ranging from 10 to more than 20 U.S. cents per kilowatt-hour. The technology is more expensive, yet it has greater potential for cost reductions in the near future (Wiser et al. 2011).

The Potential of Wind Power

In the first decade of the twenty-first century, global installed wind power capacity increased from 14 to 200 GW. At first, Germany, Spain, and the United States led the way in the use of wind power, but now the big growth is in Asia, with China alone accounting for half of all new wind power capacity in 2010 (REN21 2012). Despite the rapid expansion of wind power, it accounts for just a tiny fraction of current primary energy production (0.2 per cent), and only 2–3 per cent of global electricity production (Wiser et al. 2011).

All experts agree that the potential of wind power is far greater than current capacity, but they differ on how much greater. Estimates of global technical potential range from 70 exajoules per year (19,400 TWh per year) to 3,050 exajoules per year (840,000 TWh per year), that is, from less than 10 per cent to 600 per cent of global energy production (530 EJ).[5]

Regardless of whether the optimistic or the pessimistic estimates are more accurate, the main barrier to the expansion of wind power in the next few decades will not be the strengths of the winds themselves, but rather economic constraints (such as investments and subsidies), grid access, social acceptance and environmental impacts (Wiser et al. 2011).

4.4 Solar Thermal

As a result of the nuclear reactions occurring at its core, the sun radiates both matter and energy; matter in the form of solar winds (a stream of highly charged electrons, protons and neutrons), and energy in the form of electromagnetic radiation. The Earth's atmosphere largely shields us from the effects of solar winds, though they are felt as magnetic storms that can interfere with communications technologies, and manifest themselves visually in the auroras (Northern and Southern Lights) and the streaking tails of comets. Electromagnetic radiation, on the other hand, is keenly felt. In fact, without it there

---

[5] Estimates vary depending on whether offshore wind energy is included, on which wind speed data are used, on the areas assumed to be available for wind energy development, on the rated output of wind turbines installed per unit of land area, on their assumed technical performance, and on whether only physical limitations, or also ecological and social ones, are taken into account. Thus there is no single "correct" estimate of technical potential (Archer and Jacobson 2005; de Castro et al. 2011; Jacobson and Delucchi 2011a, 2011b; Miller et al. 2011; Wiser et al. 2011).

would be no life on our planet. It travels through space in the form of a wave, but instead of containing matter (like water in an ocean wave or air in a sound wave), these waves are composed of pure energy.

Like waves in water, electromagnetic radiation comes in varying wavelengths: radio waves are long, comparable to waves in the ocean; light waves have a medium length, comparable to ripples in a lake; and X-rays are very short, comparable to tremors in a glass of water. The shorter the length of a wave, the greater its concentration of energy. This is why prolonged exposure to certain short waves, such as X-rays and gamma rays, is so dangerous; they have sufficient energy to penetrate our bodies and interfere with our cells, causing our genetic material (DNA) to split. This, in turn, may lead to cells 'malfunctioning', forming tumours.

About 40 per cent of solar radiation is visible (what we experience as light); another 50 per cent is infrared (what we experience as heat). The rest is at higher and lower frequencies, such as gamma rays, ultraviolet, microwaves and radio waves (Arvizu et al. 2011a). Visible light is the aspect of the sun's radiation that which most crucially supports life, through photosynthesis. However, the other forms of electromagnetic radiation are also essential to life; ultraviolet radiation, for example, is used by insects to navigate, and infrared rays transmit heat.

Since any object exposed to sunlight will absorb heat, the process of converting solar energy to heat is relatively straightforward and can be achieved either actively or passively. Passive solar heating involves adapting architecture to maximise heat capture and minimise heat loss. This is the cheapest and most convenient way to take advantage of solar energy. Active solar heating involves the use of devices, usually roof-mounted panels, to collect solar radiation to heat water or other liquids.

Passive Solar Heating

The ancient Greeks were probably the first to design buildings that took advantage of the fact that, in the Northern Hemisphere, south-facing façades receive far more sunlight than those facing north, east or west do. They designed entire cities so that all homes would have south-facing façades. Later, the Romans improved on Greek solar architecture, adding two other innovations: glass windows to trap the sun's rays and black roof tiles to absorb more heat. Glass allows solar radiation to enter a building, but because in doing so it loses some of its

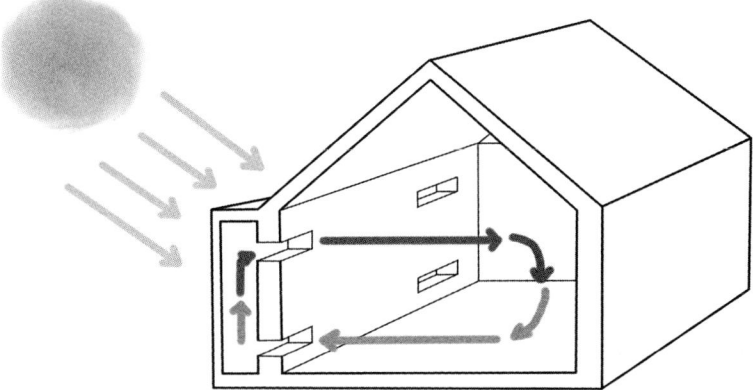

Figure 4.15. How a Trombe wall works.

energy, some of the radiation remains trapped inside. Black roof tiles heat up on a sunny day because, like all dark materials, they have the capacity to absorb electromagnetic radiation (a white surface reflects about 80 per cent of the light, a dark surface absorbs about 90 per cent). Consistent with the first law of thermodynamics (law of conservation of energy), the absorbed radiation does not disappear, but is converted into heat through the vibration of the atoms in the material.

Notwithstanding certain innovations and improvements, these three techniques – orientation, glass, and dark surfaces – remain the cornerstones of passive solar heating to this day. One of the most interesting of recent innovations is the Trombe wall, which combines all three techniques: a dark, glass-covered wall facing the sun. The solar radiation, passing through the glass panel, heats the air in the space between the glass and the black wall. Vents at the top and bottom of the wall allow a natural exchange of air through convection. By opening additional vents in the outer walls and glass panel, the Trombe wall can also be used to cool the building in summer (see Figure 4.15).

Active Solar Heating

In the early years of the twentieth century, an American entrepreneur named William J. Bailey used the evangelizing slogan "Sunshine, like salvation, is free" to sell his patented solar heating systems in California. Whereas earlier models of solar water heaters consisted of little more than a black rooftop water tank, Bailey's device separated the heating and storage elements, so that water heated during the day remained warm enough for hot showers the next morning. Bailey grew rich in a solar heating boom that ended abruptly in the 1920s as cheap diesel and natural gas became available. Being an astute businessman, Bailey made a second fortune with his Day and Night Gas Water Heater.

Modern Systems

The vast majority of solar thermal units are used to heat water to temperatures of between 20° and 80° Celsius for domestic use. In the simplest case, the unit produces warm water for sanitary use (swimming pools, showers, etc.). More complex systems are required to generate hot water for central heating. Some industrial applications also exist, in which higher temperatures are achieved. These, however, invariably involve concentrating solar radiation using parabolas and mirrors, and therefore fall within the category of concentrated solar power (see Section 4.6).

All modern solar water heating systems share two principal characteristics: the use of a collector to absorb solar radiation and a separate tank to store the hot water. The heat is transferred from the collector to the tank either using convection, the natural flow of all liquids and gases from cold to hot, or pumps that circulate the heat-absorbing fluid (see Figure 4.16).

By far the most common active solar heating systems are flat-plate collectors (covering 90 per cent of the market). These resemble solar photovoltaic (PV) panels and are generally mounted on roofs. They consist of four elements: a transparent cover that allows the sun's rays to pass through but reduces heat loss; a black sheet of metal attached to a grid of narrow tubes (the absorber); a heat-transporting fluid (usually water mixed with antifreeze) that carries the heat from the tubes to the storage tank; and an insulating backing. The fluid circulates through a grid of tubes embedded in the collector, then flows through a coil in the storage tank, releasing its heat into a tank of water that is replenished from the main water supply (see Figure 4.17).

The storage tank may either be attached to the collector (this is common in warm climates where heat loss from the tank is minimal) or located separately in a more insulated place (usually inside the building). In the case of integrated units, the storage tank is located directly above the panel, where warmer water accumulates. In separated systems the storage tank is usually located below the collector (since the collector is already located at the highest point of the building), and

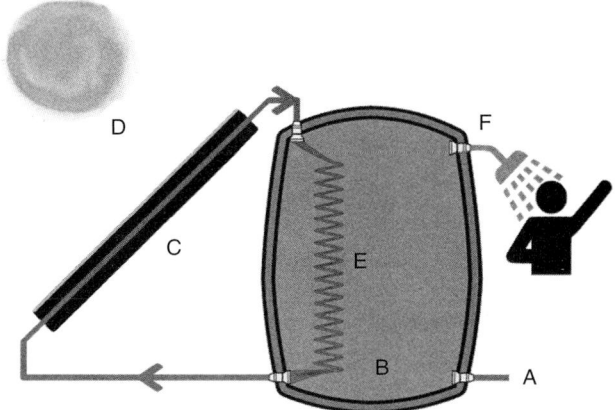

Figure 4.16. Model of a simple solar heating system. Cold water from the mains (A) flows into the storage tank, (B) a fluid passes through the solar collector, (C) is heated by solar radiation, and (D) releases its heat into the water in the tank while passing through a coil (E). The hot water rises to the top of the tank and is released through an exit valve (F).

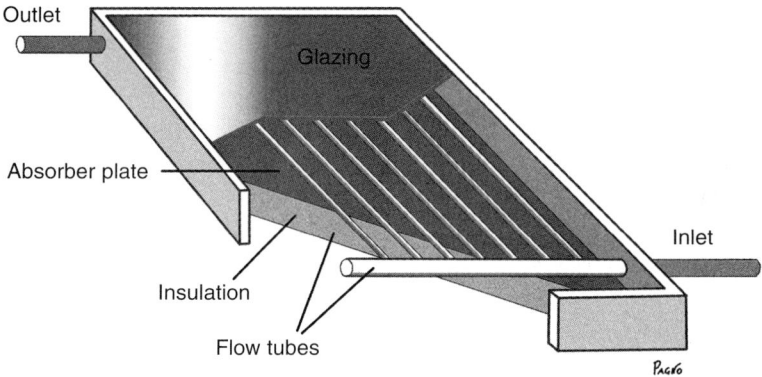

Figure 4.17. A glazed solar collector.

so it is not possible to move the warm water by natural convection. Instead, circulation is forced with the help of an electric pump.

The main alternatives to flat-plate collectors are unglazed collectors and vacuum tubes. Unglazed collectors function in much the same way as flat-plate glazed collectors, the difference being that they lack the transparent cover. This makes them considerably cheaper to produce but also less effective as they quickly lose heat once temperatures

Figure 4.18. A roof-mounted flat-plate glazed collector that is connected to a water tank inside the house. This is by far the most common technology for solar heating.
Source: SolarCoordinates at Wikipedia Commons.

drop. They are mostly used to heat swimming pools, where the desired output temperature is usually just a few degrees higher than the air temperature.

Vacuum tube collectors consist of a series of glass tubes, each containing a second light-absorbing tube. Between the two tubes is a vacuum, which allows light to pass through but prevents heat from escaping. The inner tube, which is sealed, contains a heat-transporting fluid similar to that used in the flat-plate collector (see Figure 4.19). The main difference to the flat-plate collector is that the pressure in the inner tube is very low. Pressure exerts a cohesive influence on all matter, so when it is absent or reduced, molecular bonds break more easily. As a result, the fluid vaporizes at a lower temperature than it would under normal atmospheric pressure, and travels upwards where the heat is transferred to another fluid (usually water). This secondary fluid carries heat to the house. This system is more efficient and considerably more complex and expensive than a flat-plate collector, involving different tubes and fluids to capture and transport the heat.

Benefits and Costs of Solar Heating

Up to now, solar thermal technology has mostly been used in warmer climatic zones, such as the Mediterranean, where water and spatial

How Energy Is Produced   99

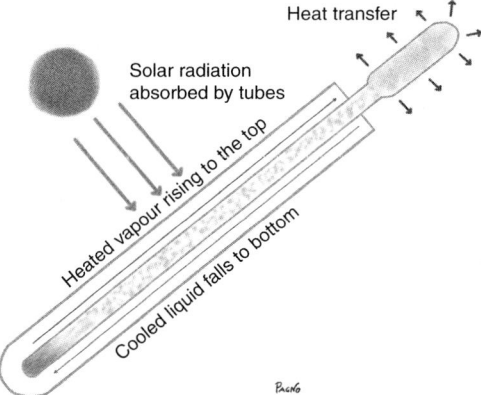

Figure 4.19. Mechanism by which a vacuum tube collector absorbs and transmits solar energy.

Figure 4.20. Vacuum tube heating system in which the water tank is integrated with the collector.
Source: Mk2010 at Wikimedia Commons.

Figure 4.21. The large-scale solar heating system that supplies hot water to the town of Marstal in Denmark, one of several pioneering projects in Scandinavia.
Source: Marstal Fjernvarme.

heating needs are modest. In cooler regions, where long, cold winters make heating an existential necessity, solar heating has not traditionally played a major role. However, as technology improves, solar heaters are becoming viable even in colder climates, but only as part of a larger energy plan and if combined with passive house technology.

Even though most solar heating systems have been installed at a scale designed to service a single household, collective facilities for apartment buildings, factories, hotels, and sports facilities are on the increase. A small number of solar thermal projects large enough to service entire towns or urban districts are currently in operation, mainly in northern Europe (see Figure 4.21). Even larger applications are in the offing. In 2011, Saudi Arabia unveiled the world's largest solar thermal plant, a 25-megawatt system that will provide hot water and spatial heating for 40,000 university students in Riyadh.[6] Since numerous industrial processes require hot water, solar heating has great potential as an industrial technology. However, in most cases higher temperatures are required than those produced by typical flat-plate or vacuum tube collectors. Higher temperatures require collectors that concentrate light using parabolas and mirrors.

Over the last thirty years, the costs associated with solar thermal technologies have fallen considerably. By now they have also reached

[6] See http://saudi-sia.com

a degree of technical maturity that makes them reliable. Passive solar technologies are already cost-effective in most locations for new buildings, but their development is limited by the slow turnover of buildings in developed countries.

Solar hot water generation costs between 10 and 20 U.S. dollars (2005) per gigajoule in countries, such as China, where the technology is widespread and therefore an economy of scale exists. It is also economical in countries with plentiful sunshine. However, in countries with little sunshine (800 kilowatt-hours per square metre per year), it may cost more than 130 U.S. dollars per gigajoule (Arvizu et al. 2011a).

The Potential of Solar Energy

The power of the sunlight hitting the Earth's surface (known as irradiance) in midday and under ideal conditions is about 1,000 watts per square metre. Considering that the sun does not shine at night, that atmospheric conditions typically reduce solar irradiance by roughly 35 per cent on clear, dry days and by about 90 per cent on overcast days, the average solar irradiance on the ground is about 200 watts per square metre. While not all countries are equally endowed with sunshine, a significant contribution to the energy mix from direct solar energy is possible for almost every country (see Table 4.4). Over one year, our planet's total solar energy 'income' has been estimated at 3.9 million exajoules (billion billion joules), more than 7 million times the primary energy supply of the human race (530 EJ). But before we grow euphoric at the potential of this largely unharnessed resource, we need to factor in a few limitations, such as land availability, meteorological conditions, local demands for energy services, and environmental and social constraints. Estimates of the global technical potential of solar energy (not diversified by technology) vary greatly, from roughly 1,500 to 50,000 exajoules per year. This is certainly cause for optimism, if not euphoria, since even the lower estimate is three times our current global supply of energy (Arvizu et al. 2011a; IEA 2012a).

The Potential of Solar Heating

More than 200 million households worldwide use solar hot water collectors. The vast majority (86.4 per cent) of glazed and unglazed water and air collectors currently in operation are installed in China (101.5

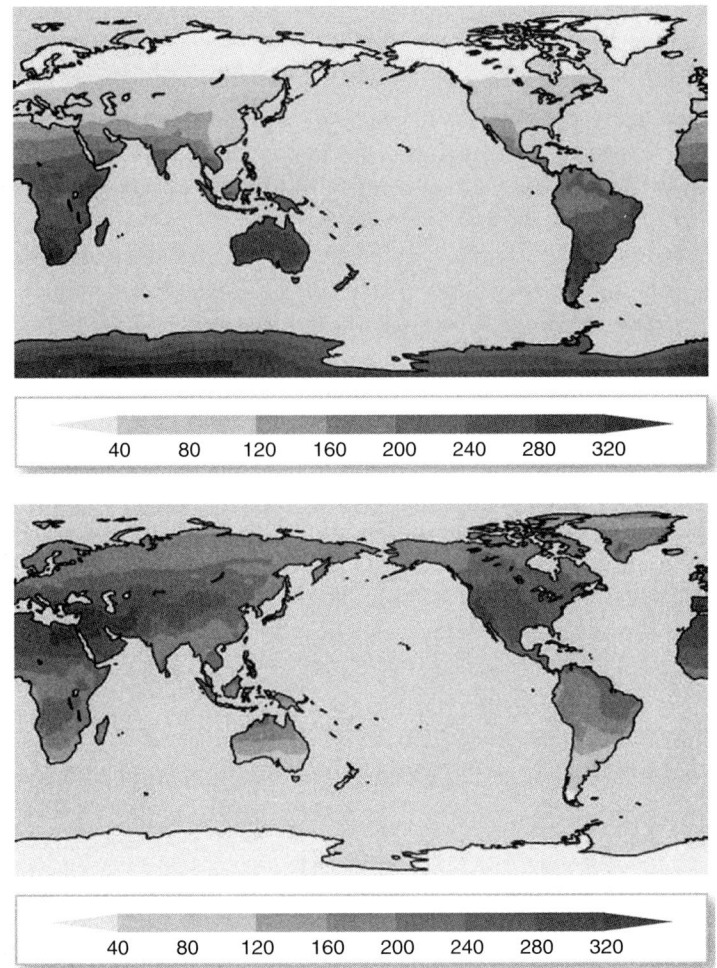

Figure 4.22. Solar irradiance (watts per square metre) on the Earth's surface in January (above) and July (below).
Source: Arvizu et al. (2011a).

GW), Europe (32.5 GW, mostly in Germany), and the United States and Canada (15 GW). Other countries in which solar water heating plays an important role include Japan, Greece, Israel, Brazil, and Austria (REN21 2012; Weiss and Mauthner 2011).

A survey of the countries that account for more than 90 per cent of the solar thermal market worldwide showed that the annual collector yield of all water-based solar thermal systems in operation in 2010 was 0.58 exajoules (162 GWh).[7] Although this roughly corresponds to 15 million tons of oil equivalent and 50 million tons of carbon dioxide savings (Weiss and Mauthner 2011), it is very low if compared with the 85 exajoules of global residential energy consumption (IEA 2012a).

Policy makers in many countries have tended to pay less attention to solar thermal technologies than to other renewable energy technologies, and the possible contribution of solar heat has been neglected in many academic and institutional energy projections. However, solar thermal technology has a considerable potential in the household, domestic, and industry sectors, and might satisfy a high proportion of energy demand in many countries, especially if combined with improved insulation and building design (see Chapter 7).

The precise technical potential of solar energy for heating purposes is difficult to assess. In a 2008 study commissioned by Dutch environmental consultants Ecofys, scientists Monique Hoogwijk and Wena Graus ventured a conservative estimate of 123 exajoules per year for solar water heating (Arvizu et al. 2011a), more than enough to meet current and future residential energy demand and much more than the meagre 0.58 exajoules that are currently supplied worldwide (see Table 4.2).

4.5 Solar Photovoltaics

The New Age of the Sun?

Most animistic and polytheistic cultures – from the Inuit to the Aborigines, from ancient Greeks to modern Hindus – have worshipped the sun. Though we have only recently grasped the scientific composition of light, our appreciation of the sun's centrality to all life is seemingly as old as humanity. The sun is the source of all life on Earth and all energy in our solar system. All plants rely on sunlight to synthesize food, and all animals rely – directly or indirectly – on those plants. Fossil fuels are ancient plant and animal biomass, wind energy relies on air movements between hot and cold zones, and hydropower depends on evaporation to raise water from oceans to mountains.

---

[7] The contribution of the total installed air collector capacity was not taken into consideration. With a share of about 0.7 per cent of the total installed collector capacity, air collectors were excluded from the calculation (Weiss and Mauthner 2011).

Table 4.2. *Estimated annual technical potential of solar energy in various regions of the world, not differentiated by conversion technology*

Regions	Range of estimates
North America	181–7,410 EJ
Latin America and Caribbean	113–3,385 EJ
Western Europe	25–914 EJ
Central and Eastern Europe	4–154 EJ
Former Soviet Union	199–8,655 EJ
Middle East and North Africa	412–11,060 EJ
Sub-Saharan Africa	372–9,528 EJ
Pacific Asia	41–994 EJ
South Asia	39–1,339 EJ
Centrally planned Asia	116–4,135 EJ
Pacific OECD	73–2,263 EJ
Total	**1,575–49,837 EJ**

Source: Arvizu et al. (2011a).

Even uranium, the fuel for nuclear power plants, is related to the sun, having been formed in the same great explosion that gave birth to most of the elements of our solar system.

Not only is the sun the origin of all energy on Earth; it is also, in itself, the most abundant of energy resources. The amount of sunlight energy that strikes the Earth in a single hour surpasses the total amount of energy consumed by humans in one year.[8] It is a truism that the future of human civilisation will depend on our ability to harness this resource since fossil fuels must, sooner or later, run out. So far we have developed two principal approaches to solar energy – collecting heat and generating electricity. Since electricity is by far the more versatile of these two energy forms (in terms of transport and conversion), it is also where the greatest hope is invested.

The first step towards converting sunlight into electricity was taken in 1839, when the French physicist Edmond Becquerel observed an increase in electrical current when the electrodes of a saline battery were exposed to sunlight. Becquerel was generally more interested in photography and optics than electricity, and he did not capitalize on

---

[8] The solar radiation intercepted annually by the Earth is estimated at 3.9 million exajoules (Section 4.4.).

his discovery. That honour fell to an American, Charles Fritts, who patented the first solid device able to convert sunlight into electricity in 1883. While Fritts's solar cell, based on the same principle as modern photovoltaic (PV) panels, represented a breakthrough, it was far too inefficient to be commercially viable, especially as far cheaper sources of electricity – fossil fuel and water-powered – were available.

It was not until the 1950s that solar electricity came into use, in an application that was appropriately futuristic: space exploration. Whereas the high research and production costs of solar panels prohibited their use on Earth, in space they represented the only viable energy source. The Vanguard satellites, America's response to Sputnik,[9] provided a compelling argument for the further development of PV panels. Instead of relying on a chemical battery that could last at most a couple of weeks, Vanguard 1 was able to transmit data from space to Earth for six years (from 1958 to 1964) thanks to solar cells mounted on the exterior. This success raised both public and scientific interest in the potential of solar power. Within a decade, the first terrestrial applications were appearing on oil rigs and pocket calculators.

How a Solar Cell Works

The development of solar PV cells coincided with the information revolution. This was not an accident, as PV cells share with all electronic devices a key component: the semiconductor. As the name suggests, semiconductors are elements or combinations of elements that conduct electricity, but only under certain conditions. Some semiconductors – such as the selenium used by Charles Fritts – are elemental. Others are synthesized by enriching natural semiconductors with other elements to magnify the semiconductor effect, a process known as doping. Whereas metals (such as iron, copper, and gold) release electrons, thereby conducting electricity, and non-metals (such as oxygen, chlorine, and iodine) adopt electrons, thus insulating against electron flow, semiconductors both conduct and insulate.

Though it sounds unspectacular, this characteristic is the lynchpin in all electronic devices, from mobile phones to the International Space Station. Because semiconductors allow us to switch the flow of electrons on and off at will, sequences of electric charge can be configured and stored. A semiconductor layer, such as that in a hard drive, is 'inscribed' with a unique combination of positive and negative

---

[9] Sputnik was the first artificial satellite, launched by the Soviet Union in 1957.

charges, which can then be 'translated', using software, into another form of information: a text, an image, computer code, or a piece of music. This effect is often explained in terms of the digits 1 and 0 (binary code).

Semiconductors are used in a slightly different way in PV cells. A PV cell consists of two layers of semiconductor material. Both layers are made of the same substance, usually silicon, but each layer is treated (or 'doped', to use the technical term) with different elements (such as arsenic, boron or phosphorus) in order to amplify either the electron-releasing or electron-absorbing effect. The electron-releasing layer, which faces the sun, is known as the n-layer (n for negative charge), while the electron-absorbing layer is referred to as the p-layer (for positive charge).

So, a PV cell consists of two layers, one inclined to release electrons, the other to receive them. But that still does not explain why the n-layer releases electrons in the first place. The answer – termed the photoelectric effect – was provided in 1905 by a twenty-six-year-old physicist working at the Swiss patent office in Bern. He showed that all matter emits electrons after absorbing electromagnetic radiation. Einstein's discovery provided a new understanding of light, proving that it is composed of discrete units, which he called quanta (and we now call photons). He explained the photoelectric effect as follows: "The body's surface layer is penetrated by energy quanta whose energy is converted at least partially into kinetic energy of the electrons" (Rigden 2005, 33).

In a green plant, photons energise the chlorophyll of the leaf, driving the conversion of carbon dioxide and water to glucose and oxygen. In a solar panel there is no chemical reaction, but the sun plays a comparable role, stimulating a flow of electrons from the silicon atoms of the n-layer. The electrons are picked up by a fine mesh of wires, and carried (via an inverter that converts direct current to alternating current[10]) to an external circuit (e.g., a lightbulb, a washing machine, or the grid). Electrons flow back from the external circuit to the p-layer. The PV cell therefore merely acts as a kind of electron pump (see Figures 4.23 and 4.24).

---

[10] In direct current (DC), the flow of electric charge is only in one direction (conventionally from a positive to a negative pole), whereas in alternating current (AC), the movement of electric charge periodically reverses direction. AC is the form in which electric power is delivered to businesses and residences. This is why a power inverter is needed.

How Energy Is Produced 107

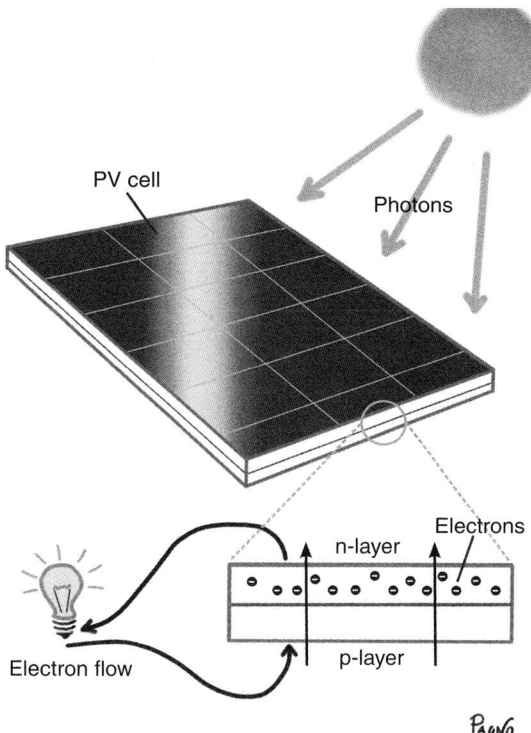

Figure 4.23. How PV cells, struck by photons, release electrons that are eventually fed into an external circuit.

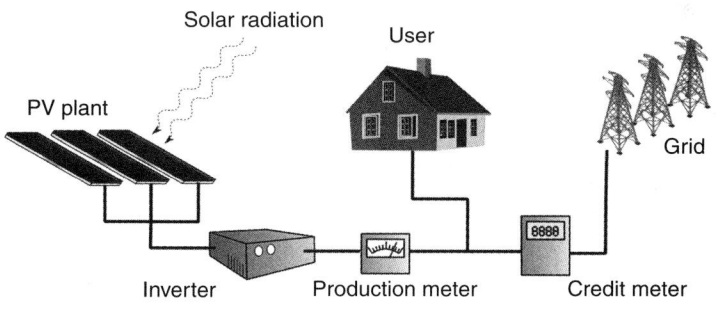

Figure 4.24. Small-scale PV plant.

The amount of current generated by a solar cell depends on the number of electrons that the photons push off the n-layer. Larger and more efficient cells, or cells exposed to more intense sunlight, will deliver more electrons.

Types of Solar PV Cells

While the vast majority of modern PV panels are made from silicon, any semiconductor material would suffice. Silicon has dominated so far because of its abundance in nature (over 90 per cent of the Earth's crust is composed of silicate minerals, combinations of silicon and oxygen) and its ubiquity in the electronics industry. The silicon used in PV cells is crystalline, meaning that its atoms are arranged in an ordered and repeated pattern. Thanks to this order, electrons can pass easily along the crystal lattice.

The most efficient PV cells are made from large crystals of silicon,[11] cut into wafer-thin slices. This is known as monocrystalline silicon. The flakes of silicon that are left over after the crystal has been sliced are used to make polycrystalline silicon (see Figure 4.25). They are heated just short of their melting point and thus 'welded' together. Since it is essentially a by-product of monocrystalline silicon, polycrystalline silicon is less expensive to produce. However, because these wafers do not have an unbroken lattice, they are less efficient in converting light to electricity than monocrystalline silicon (Table 4.3).

Even though solar PV still represents just a tiny portion of the energy mix (0.04 per cent), far lower than hydroelectricity or even wind power, solar panels have become an emblem of the transition to renewable energy. Perhaps the reason for this is their visibility. There is every chance that suburban commuters will pass several roof-mounted solar plants on their way to work, but it is less likely that they will see wind turbines or nuclear cooling towers. Roofs are the ideal surface for solar PV for three reasons. First of all, by definition the roof is the highest point of a building and therefore receives unobstructed sunlight. Second, many roofs are already inclined, allowing maximum exposure to sunlight. Third, roof mounting involves the least environmental impact because it does not usurp space from other uses. On the ground greater compromises have to be made, since solar panels require a lot of space per unit of energy generated (see Table 4.4).

[11] Bars of monocrystalline silicon are produced by dipping a 'seed crystal' into a crucible of molten silicon. In this process, a larger crystal of silicon 'grows' around the seed.

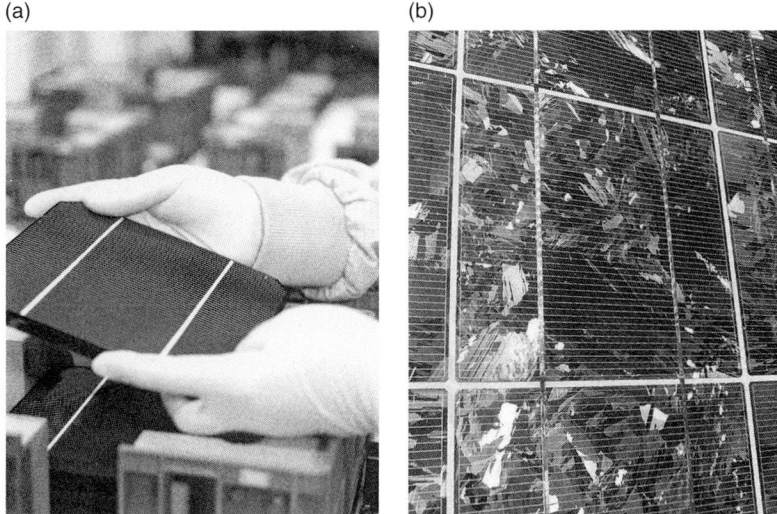

Figure 4.25. A monocrystalline silicon cell (left) ready to be assembled on a solar panel and the surface of a polycrystalline silicon cell (right). The monocrystalline cells have characteristic cutoff corners because they are sliced from a cylindrical bar of pure silicon.
Sources: Unknown at Wikimedia Commons; Georg Slickers at Wikimedia Commons.

Figure 4.26. Thin-film PV mounted on a roof.
Source: Ken Fields at Wikimedia Commons.

Table 4.3. *Cell efficiencies of different PV technologies in percentage of solar radiation converted to electricity*

Technology	Commercial	Laboratory
Monocrystalline	18–20	27.6
Polycrystalline	13–15	20.4
Thin film	8–10	20.3
Multijunction concentrator	25–30	42.4

Note: The commercial efficiency is obviously lower than records reached under laboratory conditions using experimental materials.
Source: NREL (2012).

Table 4.4. *Solar electricity that can be generated in various locations*

City	Country	Solar radiation kWh/m²/year	Usable energy kWh/m²/year	Area of PV panels m²
New York	United States	1,664	162	31
Buenos Aires	Argentina	1,840	174	29
Beijing	China	1,778	172	29
Nairobi	Kenya	1,840	174	29
New Delhi	India	2,197	197	25
Stockholm	Sweden	1,088	103	48
Cairo	Egypt	2,066	191	26
Aswan	Egypt	2,489	227	22
Venice	Italy	1,248	115	44
Sydney	Australia	1,840	174	29
Berlin	Germany	1,080	101	50

Notes: The efficiency levels of the solar PV plant are calculated at 10 per cent. The column on the right indicates the area needed to produce 5,000 kilowatt-hours per year, the electricity consumption of a typical European household.
Source: www.nrel.gov.

Thin flexible film that can be stretched over almost any surface may help to resolve this conflict. This removes the need to mount the cells on a rigid frame and opens up a range of futuristic possibilities, including solar cars and solar clothing. However, thin-film cells have significantly lower conversion efficiencies than silicon does. To

Figure 4.27. Concentrated PV plant. Most of the surface of the panel is occupied by optical concentrators, thousands of small lenses, each focusing roughly 500 times the sunlight onto a tiny high-efficiency multijunction photovoltaic cell. A Tesla Roadster Electric Vehicle is parked beneath for scale.
Source: Mbudzi at Wikimedia Commons.

compensate for this, several different films may be layered to create what are known as multijunction PV cells. In this case, each layer or junction (the word 'junction' is preferred since each cell already consists of two layers, the p-layer and the n-layer) uses a different type of semiconductor, each of which absorbs light from a different part of the colour spectrum. In this way, the multijunction cell is able to maximise energy gain with a typical conversion efficiency of 30 per cent. The downside of this technology is that it is more complex and therefore expensive to produce. This disadvantage may be offset by concentrating the sunlight using mirrors or lenses before it strikes the cells in a technique known as concentrated PV (CPV). CPV systems are considerably more efficient than flat silicon panels, thus requiring less PV material to produce the same amount of electricity (see Figure 4.27).[12]

---

[12] Multijunction concentrators have achieved more than 40 per cent conversion efficiency in laboratory tests. To maximise efficiency, CPV systems use solar tracking (the array follows the path of the sun through the sky) and must be located in areas that receive plentiful direct sunlight. Diffuse light, which occurs in cloudy and overcast conditions, cannot be concentrated.

Figure 4.28. An experimental solar airplane, built by NASA, and the International Space Station, whose solar panels provide 110 kilowatts of power, roughly equivalent to the requirements of fifty-five American households.
Source: NASA (www.nasa.gov).

From Cells to Plants

By the early 1960s, solar PV technology had advanced to the point where conversion efficiencies were almost on par with modern commercial panels. Yet it took another three decades for solar cells to be produced on an industrial scale. The reason for the delay was cost. Solar cells are expensive to manufacture because the silicon they contain must be refined to a purity level of at least 99.9999 per cent in order to function effectively. So onerous are the research, development, and

Figure 4.29. One of the authors installing a 4.7-kilowatt PV plant on the roof of his house.
Source: Gian Andrea Pagnoni.

production costs of solar technology that, like nuclear energy, it would probably never have emerged without major state investment. This investment was forthcoming thanks first to the space race, then to the oil crises and, more recently, to the drive for cleaner and more sustainable energy.

PV cells require protection from the elements and are therefore usually encased in glass. When more power is required than a single cell can deliver, cells are connected to form PV modules. Modules are the fundamental building block of PV systems. A single module is enough to power certain applications, such as an emergency roadside telephone, but for a house or a power plant the modules must be arranged in multiples as panels and arrays. Panels are pre-wired and ready to install. The modular design of a PV panel is one of its greatest advantages. As demand grows or shrinks, a plant can be easily expanded or downsized. It is also possible to mix panels from different manufacturers, provided they have the same voltage output. A PV array consists of a number of individual PV panels that have been wired together in a series. An array can be as small as a single pair of panels or large enough to cover several hectares.

Figure 4.30. A vast array of solar panels at Serpa Solar Park in Portugal.
Source: Ceinturion at Wikimedia Commons.

Figure 4.31. BIPV (Building Integrated Photovoltaic) are PV materials that are used to replace conventional building materials in parts of the building, such as the roof, skylights, or facades. This sports hall in Germany is covered with PV panels.
Source: Björn Appel at Wikimedia Commons.

How Energy Is Produced   115

Figure 4.32. A single solar PV module used to power a roadside emergency telephone in France.
Source: Gian Andrea Pagnoni.

Figure 4.33. A small solar thermal plant and an 87-kilowatt off-grid PV plant in rural India.
Source: Gian Andrea Pagnoni.

The Benefits and Costs of Photovoltaics

Between 1995 and 2010, global PV capacity increased more than eightyfold, from 0.5 gigawatts to approximately 40 gigawatts (REN21 2011, 2012), making PV the world's fastest-growing energy technology. Most of this additional capacity has been installed in Europe (with almost a third in Germany alone), North America, and China. But India is also rapidly expanding its solar PV sector with giant new facilities such as the Charanka Solar Park in Gujarat (500 MW).

At the other end of the scale, there has also been a proliferation in recent years of small stand-alone PV applications. These vary in size from wristwatches and calculators to remote buildings, ships and spacecraft. These systems are usually combined with a battery to ensure power supply at night and in overcast conditions.

PV solar plants are ideally suited to the regions of the world with the highest population densities, and they are viable at most latitudes. Because of their modular nature, plants can start small and be gradually expanded as demand increases or economics allow. The main limitations of PV plants are intermittency and costs. The problem of intermittency is best resolved by combining PV with other energy sources, such as wind power, hydropower, or biofuels. Costs per kilowatt-hour of PV power have fallen dramatically in the last twenty years but remain higher than for most other renewable energy sources (see Section 6.14).[13]

As the scale of production increases, and efficiencies increase, costs will fall further, but it is unlikely that PV will be viable without state incentives and subsidies in the near future. Because the process of producing pure silicon is energy intensive, and many of the elements used to dope the cells are rare, the material costs of PV will remain relatively high. Assuming the PV market continues to grow at more than 35 per cent per year, the cost of PV-generated electricity is likely to drop by more than 50 per cent to about 7.3 U.S. cents (2005) per kilowatt-hour by 2020 (Arvizu et al. 2011b).

The holy grail for renewable energy is a factor known as 'dynamic grid parity'. This is achieved when, in a particular market, the cost of producing power from a given energy source is equal to the

---

[13] Average global PV module factory prices dropped from about 22 U.S. dollars per watt in 1980 to less than 1.5 U.S. dollars per watt in 2010. PV electricity generation costs range from 14 to 36 U.S. cents per kilowatt-hour. In regions of high solar irradiance in Europe and the United States (> 1,800 kilowatt-hours per square metre per year) costs are in the range of 19–22 U.S. cents.

cost of buying electricity from the grid. According to researchers at Padua University (Italy), solar PV will soon be cost-competitive in most parts of Europe. PV grid parity is expected to be achieved first in the industrial sector and in the residential sector between 2015 and 2020 (Cavallin et al. 2011; Didier 2011; Meneguzzo 2011).

However, solar PV is still highly dependent on state subsidies and supports. These come in two main forms: (1) laws that oblige electricity utilities to produce a specified fraction of their electricity from renewable sources and (2) price supports, guaranteeing a certain price for renewable electricity sold to the grid. The aim of these supports is to assist the renewable sector, particularly wind and solar power, to reach a scale of production that is competitive with conventional power production. According to a recent study by the University of Arizona (Gowrisankaran et al. 2011), if the share of PV in the U.S. electricity mix were to rise to 20 per cent, the average cost per kilowatt-hour would be less than 14 U.S. cents. Moreover, if externalities such as carbon dioxide emissions were factored into the production costs of electricity, solar power would immediately become more competitive with fossil fuelled power. This would, in turn, promote further expansion of the solar sector, leading to greater economies of scale and reduced costs per watt.

Italy and Germany are two countries that have implemented generous price supports, known as feed-in tariffs, for solar PVs. This has led to dramatic growth in the PV sector in both countries. From 2000, when Germany first introduced feed-in tariffs, to 2012, the country's PV capacity increased from 113 to 32,411 megawatts. In Italy, the growth of the PV sector has been even faster. Just five years after the introduction of feed-in tariffs in 2007, Italy's PV capacity had grown from 120 to more than 16,000 megawatts (Eurobserver 2013). This has had several consequences in the two countries. It has raised the prospect of greater energy independence at a time when both countries' external supply of fossil fuels is unstable (Germany relies heavily on Russian gas, while Italy has imported a lot of its oil from Iran). In Germany's case it has also bolstered the ambitious national commitment to an 'energy transition' from fossil and nuclear energy to renewables.

The growth of solar PV has also had an unexpected positive effect on the entire electricity system in both countries, smoothing the peaks of electricity demand during midday. Before the PV boom in Italy, there were two price peaks in the power market, one during the day (around 11:00 AM) and another in the evening (between 6:00

and 8:00 PM). The eleven o'clock peak has now almost disappeared, as during central daytime hours PV competes with fossil fuel–based power stations, restraining the price of electricity (Energia24 2012; QualEnergia 2013a).

How Many Panels Would Each Roof Need?

The market conditions for PV varies considerably from country to country. This is attributable to three main factors: differences in state energy policies, differences in public enthusiasm for renewable energy, and differences in the amount of available sunlight. As with wind power, some regions are far more suited to solar energy than others. Unfortunately, those countries with the highest levels of industrialisation and the greatest energy demand tend not to be those with the highest incidence of solar radiation. Average solar radiation varies from about 250 watts per square metre at hot spots such as the Sahara Desert, the Arabian Peninsula, and central Australia to about 50 watts per square metre near the poles.

The average efficiency value (from sun to electricity) of solar PV cells is just 10 per cent. This is because most PV panels have conversion efficiencies of 20 per cent or less, some energy is lost through inverters and cables, and about 15 per cent of the area of a panel is not covered by solar cells. In Germany, for example, the average solar radiation measured over a period of a year (including nighttime) is about 150 watts per square metre. We can thus calculate an annual energy output of 131 kilowatt-hours per square metre per year. Considering that the average German household consumes about 5,000 kilowatt-hours of electricity per year, that household would require 38 square metres of solar panels to meet its electricity needs. The productivity of solar PV systems also depends on a few other factors, most importantly the orientation of the panels (the optimal tilt depends on latitude and time of year) and shading from trees, chimneys or other objects.

The Potential of Solar Photovoltaics

In the last ten years, the solar PV market has seen extraordinary growth, from 1.5 GW of installed capacity in 2000 to 70 GW in 2011. The top countries for total installed capacity in 2011 were Germany, Italy, Japan, and Spain, followed closely by the United States, China, Japan and Australia. The leaders for solar PV per inhabitant were all in Europe: Germany, Italy, the Czech Republic, Belgium, and Spain (REN21 2012).

Most installed PV capacity today is grid-connected, and small-scale off-grid applications account for a declining share of overall capacity. This is linked to the fact that most of the incentives offered for PV installation are available only for grid-connected systems. However, with time and as new funding models emerge, off-grid solar PV is likely to expand, particularly in developing countries or remote locations that lack an extensive grid network. Just as mobile phone technology has revolutionised communications in many parts of Africa, Asia, and South America, small-scale off-grid solar PV plants have the potential to bring electrical power to communities that have lacked electrical power up to now.

Estimates for the annual total technical potential of solar energy (not differentiated by conversion technology) vary widely between different studies, from roughly 1500 to 50,000 exajoules per year (Arvizu et al. 2011a; IEA 2012a). The main difference between the studies arises from assumptions about the power conversion efficiency of the technology and the allocated land area availabilities. A recent comparative study estimated future technical potentials of photovoltaics at roughly 1,700 exajoules per year in 2050 for photovoltaics (Arvizu et al. 2011a; Krewitt et al. 2009). This represents more than three times the current global primary energy production (530 EJ).

4.6 Concentrating Solar Power

Archimedes' Mirrors

In 212 BCE, Syracuse, the last of the Hellenic colonies in Italy, fell to the Romans after a long siege. That this city held out for sixty years after the rest of Magna Graecia had capitulated was largely thanks to its excellent defences, designed in part by the inventor and mathematician Archimedes. Among the many impressive weapons he is reported to have devised were giant mirrors that concentrated sunlight on the Roman ships, causing them to burst spontaneously into flame. Early modern scientists, including Leonardo da Vinci, Descartes and Kepler, were both fascinated by and sceptical of the accounts of Archimedes' mirrors,[14] and modern reconstructions have cast further doubt on the feasibility of such weaponry with the limited means

---

[14] Their principal source was the second-century satirist Lucian.

available to Archimedes. Had the discoverer of the principle of floatation not perished on a Roman sword at the end of the siege, perhaps we would know more today about his 'burning mirrors'.[15] Whatever its veracity, this story advanced the idea that the sun's rays, if concentrated, were a powerful source of energy. It was not, however, until the second half of the nineteenth century that a device capable of effectively harnessing that energy was developed.

Solar Ice

Augustin Mouchot was convinced that coal supplies would eventually run out, and set out to discover an alternative source of energy. At the 1878 World Exposition in Paris, he unveiled a miraculous device: a machine that converted the heat of the sun into blocks of ice. Mouchot's parabolic solar concentrator produced enough heat to drive a steam engine, which, in turn, powered an ice maker.

Despite the great interest aroused by Mouchot's invention, his engines never passed the experimental stage. He returned to teaching mathematics, while his assistant, Abel Pifre, went on to design several other solar-powered machines (see Figure 4.34). However, because of their low efficiency and the comparative cheapness of coal, investors were hard to find. Over the course of the nineteenth century, numerous patents were registered for devices that took advantage of concentrated solar power (CSP), but investor interest invariably fluctuated with the price of fossil fuels. This principle has held true to this day; recent interest in CSP was sparked by the oil crises of the 1970s.

Power from the Sun: How CSP Works

When I was eight, my father showed me how to singe dry leaves with the lenses of my grandfather's glasses. Imagining myself as Ra, the Egyptian sun god, I spent the afternoon subjugating our back garden. Each leaf was a disobedient city rebelling against my authority. Unfortunately, the summers in Italy's Po Valley are wet as well as hot, and this put a dampener on my tyrannical ambitions. The concentrated

---

[15] Since the 1970s there have been a number of attempts to replicate Archimedes's deadly heat ray, including a 2005 experiment at the Massachusetts Institute of Technology. It was found that the bronze mirrors available at that time could hardly have ignited vessels hundreds of metres from Syracuse's walls (Zamparelli 2005a, 2005b).

Figure 4.34. A solar-powered printing press invented by Abel Pifre at the end of nineteenth century.

rays from Beppino's glasses, far from producing the desired inferno of foliage, merely peppered some damp leaves with brownish holes.

Modern CSP power plants are not terribly different – in terms of the design principle – from those developed by Mouchot and Pifre. A collection of mirrors focus the sun's rays onto a target point, where the heat can be converted into electricity. Several different technologies are available, but the most common is based on parabolic troughs. Shaped like gutter piping, the trough is designed to automatically follow the path of the sun through the sky, thus ensuring that it receives maximum direct sunlight throughout the day (see Figures 4.35 and 4.36). The mirrors that line the trough concentrate sunlight onto receiver tubes that contain a fluid, such as synthetic oil, that can be heated to high temperatures (approximately 400°C). As in a solar thermal plant, this oil, known as a thermal transfer fluid because it allows heat to be transmitted, is pumped through a series of heat exchangers to produce steam, which in turn drives a turbine to generate electricity.

The most common alternative designs are solar towers and parabolic dishes (see Figure 4.35). A solar tower is, as the name suggests, a tower with a barrel-like fluid container (the solar receiver) mounted at its apex. Surrounding the tower in a fanlike array are numerous large mirrors with sun-tracking motors, known as heliostats (see

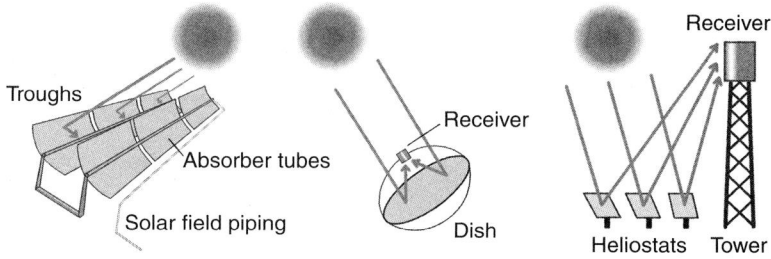

Figure 4.35. The three main models of CSP plant: linear trough concentrator, parabolic dish, and solar tower.

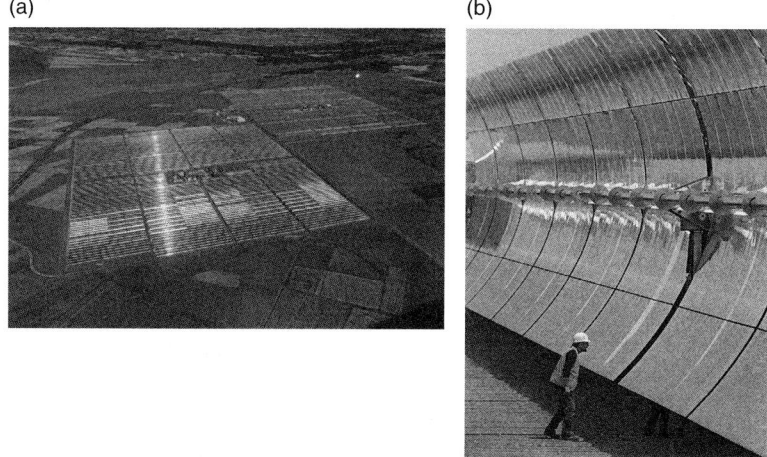

Figure 4.36. Linear trough concentrators at the Andasol Power Station (Spain).
Source: Solar Millennium.

Figure 4.37). Because so many mirrors concentrate sunlight onto a single point, it produces precisely the effect that Archimedes set out to achieve, with temperatures exceeding 1,000 degrees Celsius. As with the parabolic trough, the heat is delivered to a steam turbine.

A parabolic dish resembles a TV satellite dish; an umbrella-shaped reflector tracks the sun and concentrates light onto a receiver located at the focal point. The receiver heats a fluid or gas, which is used to generate electricity using a heat-driven engine or micro-turbine attached to the receiver (Richter et al. 2009; JRC 2011). The major advantage of the parabolic dish over the other models of CSP plant is that it can be installed on a very small scale, since each dish operates as a stand-alone unit, similar to a wind turbine.

Figure 4.37. The Solucar PS10 CSP power station in Seville, Spain, uses hundreds of flat mirrors to reflect light onto the receiver. It is the first commercial solar thermal power plant based on a tower.
Source: SOLUCAR PS10/afloresm at Wikimedia Commons.

Coping with Intermittency

CSP technology is already capable of producing electricity at an industrial scale. As with PV and wind, the main problem is intermittency. Moreover, CSP plants can only operate in hot, bright, cloudless conditions. This makes them viable in only a narrow band of climatic zones, and even there only during the day. Thermal storage offers a way around this problem. During the day, some of the heat generated by the CSP plant is used for power production, while some is diverted to a storage material, usually molten salts in insulated tanks (see Figure 4.38). These salts fulfil the same function as the fluid that runs through a linear trough concentrator, except that their main function is to store rather than to transport heat. Because salts can be heated to far higher temperatures than oils (up to 550°C), they are more suitable for stationary storage, though less suitable for pumping through pipes (IEA 2010b; JRC 2011). At night, the heat stored in the molten salts can be used to produce steam and thus electricity.

The Benefits and Costs of CSP

CSP can provide carbon-free renewable energy in countries or regions with strong sunshine and clear skies. It can produce significant

Figure 4.38. Insulated storage tanks of molten salts at Andasol Power Station (Spain).
Source: Solar Millennium.

amounts of high-temperature heat for industrial processes, and peak load power during daytime. Furthermore, in arid countries it can meet growing demand for water desalination (DESERTEC 2009; IEA 2010b).

Parabolic troughs are the most mature of the CSP technologies and therefore dominate the current market (REN21 2012). Currently, the cost of installing a 50-megawatt CSP plant is around €4,800 per kilowatt. The costs of electricity from large solar troughs with six hours of thermal storage range from 20 to 30 U.S. cents (2005) per kilowatt-hour. In very sunny locations, with solar radiance of 2,500 kilowatt-hour per square metre per year (e.g., in North Africa, Middle East, the southern United States, and central Australia), costs may be 20–30 per cent less (Arvizu et al. 2011a; IEA 2009a; JRC 2011; REN21 2012).

Although global capacity increased almost fivefold, from 354 to 1,707 megawatts between 2005 and 2011, CSP has not seen the same explosive proliferation as solar PV or wind power have. When we consider that 95 per cent of that capacity is in just two countries (Spain and the United States), it is clear that CSP as a global energy solution is still in its infancy. Currently, the deployment of CSP technology is limited by regional availability of good-quality irradiance of 2,000 kilowatt-hours per square metre or more (Arzivu et al. 2011a). New projects are under way in Italy, the Middle East, North Africa, India, and China (IEA 2009a; JRC 2011; REN21 2012; Shahan 2012).

The Potential of CSP

Carlo Rubbia, the Italian nuclear physicist and Nobel laureate, has long been a strong advocate of CSP. In 2005, after completing his term as director of ENEA,[16] he was appointed principal scientific adviser to CIEMAT, a Spanish public research organization on energy, environment, and technology. Under his guidance, Spain has now become the world leader in CSP (Ombello 2010). According to Rubbia (2012), the technical potential of CSP, especially in the vast desert zones, is immense but barely tapped. Incredible as it may sound, the amount of solar energy that strikes the world's deserts in six hours is roughly equal to humankind's entire annual energy consumption. Roughly 1 per cent of the 36 million square kilometres of the Earth's deserts, an area roughly the size of Germany or Japan, would therefore be sufficient to satisfy the entire annual primary energy consumption of humankind (DESERTEC 2009).[17]

CSP potential is estimated at roughly 8,000 exajoules per year, 15 times current global production (Krewitt et al. 2009). Therefore, the main limitation to CSP expansion is not the availability of suitable ground space, but rather technical and economic factors. Technical limitations include the fact that, as an intermittent energy source, CSP cannot provide base-load power as fossil fuels do. Also, because CSP plants are best suited to hot deserts, long distances between the sites of production and consumption need to be bridged. The idea of generating solar energy in deserts and transporting it to urban areas has been promoted, in particular, by the DESERTEC Foundation, a network of scientists and business interests whose long-term goal is to meet most of the energy needs of North Africa, and as much as 10 per cent of Europe's, through solar power plants installed in the Sahara (DESERTEC 2009; IEA 2010b).

CSP-generated electricity is still expensive compared with fossil-fuelled power, and this remains a major barrier to development. Somewhat ironically, dramatic reductions in the installation costs of PV plants are challenging the CSP market in the United States and Europe. In the last few years, several planned CSP projects have been redesigned to use PV technologies instead. According to some experts,

---

[16] ENEA is Italy's National Agency for New Technologies, Energy and Sustainable Development (www.enea.it).
[17] Diverging figures have been provided by other studies, which suggest that 15 per cent of the 15 million square kilometres of deserts will be necessary to meet global primary production (Andrews et al. 2011; McDonald et al. 2009).

the rush towards PV may represent a growing trend, while others believe that the provision of thermal storage will justify a moderately higher price for CSP (REN21 2012). Because CSP is still in the research and development phase, its future role in the global energy mix is uncertain.

4.7 Bioenergy

The Sun in a Teacup

The sugar that sweetened my tea this morning was a kind of stored solar energy, formed months ago by plants. Although most of the solar energy that penetrates the Earth's atmosphere is expended in warming the soil, air, and water, and in driving winds and waves, a tiny fraction (about 0.1 per cent) is exploited by plants and algae to fuel photosynthesis, in which atmospheric carbon dioxide and water are converted into glucose (a simple form of sugar) and oxygen. Once produced, this glucose may take one of three different paths: either it is consumed immediately within the plant's cells to provide energy for metabolism; it is converted into sucrose or starch and stored as an energy reserve within roots, fruits, or tubers; or it is converted into the cellulose and lignin that form the plant's skeleton.

When plants and animals die, their biomass is decomposed by bacteria, the complex organic molecules broken down into inorganic carbon dioxide and water, which re-enter the photosynthetic cycle. 'Biomass' is the term given to all organic matter produced by the organisms that inhabit the biosphere.[18] Most biomass circulates continuously through the food chain, but a part may be stored for a time in the form of peat or fossil fuels. Fossil fuels have an organic origin but are not considered biomass because they have left the photosynthetic cycle. This is an important distinction in energy terms. Biomass fuels, as long as they are produced sustainably, avoiding forest destruction and lengthy transportation, are carbon-neutral, since the carbon dioxide released through combustion in a stove or power plant is recaptured through photosynthesis to build new biomass.

---

[18] The biosphere is the relatively thin layer that extends from the ocean floor to an altitude of about 10 kilometres above the Earth. As the term suggests, it is the sphere that supports life.

Table 4.5. *The energy content of various fuels*

Fuel	GJ/ton
Fresh grass	4
Green wood (50–60% humidity)	6–7
Municipal solid wastes	9
Air dried wood (30%–20% humidity)	14–18
Straw, bagasse	15–18
Dry dung	8–16
Paper from newspapers	17
Biogas	12–20
Lignite	10–25
Coal	28–31
Charcoal	28–30
Bioethanol	30
Biodiesel EN14214	37.2
Rapeseed oil	37–38
Sunflower oil	38
Diesel	42
Crude oil	42
Natural gas	43–55
LPG	46

Extracting Energy from Biomass

Biomass can be divided into two broad categories: traditional and modern. Traditional biomass includes wood, charcoal, agricultural residues and animal dung used for cooking and heating, mostly in poorer communities. Modern biomass encompasses biomass crops, biogas, wastes, forest and food industry residues for electricity and heat generation, together with fuels (such as bioethanol and biodiesel) for transportation (IEA 2012a).

Biomass may take solid, liquid, or gas form, and there is a wide variance in terms of energy content; from fuel wood that contains about 15 GJ of energy per tonne, to sunflower oil at 38 GJ per ton, not much less than petroleum (see Table 4.5). Biomass products may be divided into four main categories:

1. **Solid biomass**, sold as wood logs, chips, pellets, or briquettes, is usually derived from crops such as willow, poplar, eucalyptus

and miscanthus that can be planted and harvested in short rotation.[19]
2. **Wastes** such as sewage, sludge or slurry; solid domestic and commercial waste; and residues from agriculture and industries, such as timber, straw, paper pulp and fruit stones.
3. **Biogas**, a mixture of carbon dioxide and methane that is produced when bacteria digest certain low-fibre biomass.
4. **Liquid biofuels**, such as bioethanol or biodiesel, which are produced by fermenting sugar and refining vegetable oils.

Burning Solid Biomass

Anyone who has been on a camping trip will have noticed that it takes a lot of foraging to gather enough firewood to make dinner. Boiling a pot of water on an open fire can be a very satisfying experience, but as an energy conversion process it is extremely inefficient. Under ideal conditions (100 per cent efficiency) 37 cubic centimetres of wood (just one or two twigs) is all that is needed to raise the temperature of a litre of water from 20°C to 100°C. However, a traditional stove, with a typical conversion efficiency of 25 per cent, requires about four times as much wood, an open fireplace would consume ten times, and a campfire perhaps twenty times as much.

Technologies for biomass combustion range from domestic stoves and boilers to sophisticated plants that heat entire districts or generate electricity. A power plant fuelled with solid biomass works in the same way as a coal-fired power plant, except that it burns wood pellets or dry wastes to produce heat or electricity, or both. The biggest difference between biomass and coal plants is in scale. Biomass applications range in output from a few kilowatts for small residential applications (stoves, boilers, and fireplaces), to several megawatts for district heating systems, and up to 150 MW for biomass power plants. These plants are dwarfed by the average coal-fired power plant with a capacity of 1,500 megawatts. The use of biomass alongside coal is currently the most cost-effective and energy-efficient way to produce heat and electricity from biomass.

---

[19] Conventional forestry involves cultivating trees until they have grown large enough to be commercially mature. Depending on the species of tree and the quality of the land, this takes between ten and forty years. Short rotation forestry refers to trees in very dense plantations, harvested at intervals of three to eight years to provide a regular and renewable supply of fuel.

Figure 4.39. Short-rotation poplar plantation in Italy (to produce wood for combustion in a solid biomass power plant). The plantation is fertilized with digestate from a biogas power plant.
Source: ASICOOP Srl.

Charcoal and Pellets: Energy-Dense Solid Biomass

The value of woody biomass as a fuel depends on its density and moisture content. Wood with a moisture content above 67 per cent will not burn. Freshly felled timber, which has a moisture content of 50–60 per cent, burns poorly, while wood that has been left to dry for a year, with a typical moisture content below 30 per cent, burns well. The most combustible of all wood products is charcoal; its moisture content is negligible, its energy density is similar to coal, and when burnt it produces very little smoke and soot. This explains why charcoal was the industrial fuel of choice until the nineteenth century, used for firing bricks, making lime and smelting metal. Charcoal is produced by burning wood in a low-oxygen environment. This is traditionally done by covering a wood pyre with clay, leaving small air holes at the bottom. Wood, like all biomass, is composed mainly of carbon, hydrogen and oxygen. The heat expels most of the hydrogen and oxygen in the form of water vapour, leaving almost pure carbon with an energy density almost twice that of the original wood.

Figure 4.40. A 10-kilowatt modern pellet stove. Thanks to the convenience and high calorific value of pellets, combined with government incentives, this technology has taken off in many Western countries.
Source: La Nordica Extraflame.

Charcoal is therefore an energy-dense, clean and convenient wood-derived product. Today, its most common use is in barbeques, rather than industry. A more modern densified form of bioenergy is wood pellets, which are produced by compressing dried sawdust. Because of their low moisture content and higher density, pellets have a much higher calorific value than wood. Pellets are used instead of logs in modern domestic stoves and boilers (see Figure 4.40), and are already competitive with fossil fuels in terms of price.

Copying Cows: Biogas

Two billion years ago, before algae and plants boosted the oxygen content of the atmosphere through photosynthesis, anaerobic bacteria were the most common form of life on Earth (Biello 2009). They are still around in force, digesting organic matter in airless conditions, thriving in swamps, oil fields, rice paddies, landfill sites, and in the guts of ruminant mammals.

Figure 4.41. An 80-megawatt solid biomass power plant in Italy. Heaps of fuel (wood chips and timber industry residues) are stored next to the plant. The wood chips are fed into the boilers, and electricity is produced using a traditional steam turbine. If suitably located, the plant may also act as a source of district heating.
Source: Gian Andrea Pagnoni.

A cow happily grazing on a meadow seems perfectly self-contained, yet without the help of an army of bacterial accomplices she would be unable to digest a single blade of grass. Ligno-cellulose fibres, the material that gives plants their physical structure, consist of rigid lattices of molecules that provide support for vertical growth and stubbornly resist biological breakdown (Huber and Dale 2009). Ruminants, including cattle, sheep and deer, have a specialised digestive system that allows them to break down this fibrous biomass into digestible sugars. A ruminant's digestive system includes a special chamber, known as the rumen, where complex plant matter such as cellulose is fermented with the aid of anaerobic bacteria. A by-product of this specialized digestive process is biogas, mainly consisting of methane (60 per cent) and carbon dioxide (40 per cent). Traces of other gases, such as hydrogen, carbon monoxide, nitrogen, oxygen and hydrogen sulphide, are also present. Although humans are known to occasionally produce and release methane,

Table 4.6. *Methane yields from digestion of various plant materials*

Feedstock	Crop yield t/ha	CH$_4$ yield m^3/t	CH$_4$ yield m^3/ha
Maize (whole crop)	9–30	205–450	1,660–12,150
Wheat (grain)	3.6–11.7	385–425	1,245–4,500
Barley	3.6–4.1	350–660	1,445–2,430
Grass	10–15	300–470	2,680–6,300
Miscanthus	8–25	180–220	1,290–4,900
Sunflower	6–8	155–400	830–2,880
Oilseed rape	2.5–7.8	240–340	540–2,390
Peas	3.7–4.7	390	1,300–1,650
Potatoes	10.7–50	270–400	2,660–18,000
Sugar beet	9.2–18.4	235–380	1,954–6,310

Note: High net energy yield per hectare is an indispensable prerequisite for an economic operation of a digestion plant.
Sources: Braun et al. (2009), Murphy et al. (2011).

this is associated with indigestion and is not a standard feature of our digestive process.

If we place wet biomass in a warm closed tank, we reproduce the conditions in a cow's intestine. So far, biogas tanks are far less efficient than their natural precursors. The best artificial anaerobic digesters convert biomass to biogas at a rate three times lower than a cow's intestine (and fifty times lower than a termite's!). Moreover, up to now it has not been possible to effectively breed bacteria that digest lignin and cellulose, which is why biogas plants currently rely on wet biomass from agriculture (e.g., maize, sorghum, potatoes), municipal and industrial organic wastes, sewage, slurry, and slaughterhouse waste.

Artificially produced biogas can be used in much the same way as natural gas. A biogas power plant consists of four sections. The raw fuel (known as feedstock) is stored in trenches and silos. This is fed into a digester, a sealed tank where bacteria break down the organic matter in a fermentation reaction to produce biogas. This biogas is first treated to reduce humidity, and to remove sulphur and other trace pollutants, then piped to the power station. Here the

How Energy Is Produced   133

Figure 4.42. Maize and food waste such as potatoes and tomato skins for anaerobic digestion. In most biogas power plants slurry or sewage is added to aid digestion.
Source: Gian Andrea Pagnoni.

Figure 4.43. Anaerobic digesters at a 1-megawatt biogas plant, the scale currently preferred in Europe. Inside the tanks, a temperature of 30–50 degrees Celsius allows bacteria to thrive and digest biomass, releasing biogas. The dark building houses the generator.
Source: Gian Andrea Pagnoni.

Figure 4.44. The biogas produced in the digester is purified, then conveyed to the power station, where a 4-stroke engine (right) is connected to a generator (left).
Source: Gian Andrea Pagnoni.

biogas fuels an internal combustion engine. Obviously, the energy output of biogas is much lower than that of natural gas since half of it comprises carbon dioxide, which doesn't burn. The engine shaft is connected to a generator which produces electricity (see Figure 4.44). Some of the heat produced through combustion is pumped back into the digestion tank to maintain the warm temperature. Digested material (known as digestate) is regularly removed from the tank. Properly treated, this is a valuable agricultural fertilizer (Al Seadi and Lukehurst 2012; Braun et al. 2009; Murphy et al. 2011; Petersson and Wellinger 2009).

Heat and Power from Wastes

For the last 10,000 years, cultivated crops have given us food and fodder, construction materials, and fibres for clothing and paper. Agriculture has also produced large quantities of waste (primary residues), such as stalks, branches, and animal dung. Secondary residues arise after the product has been used, and include nutshells, tomato skins, demolition timber, sawdust, sewage, and municipal wastes.[20] Both primary and secondary residues can be used, after proper treatment, as biomass feedstocks. With dry or solid waste (such as wood shavings) the preferred route to energy generation is direct combustion; with wet and low-fibre biomass (such as organic municipal wastes), it is anaerobic digestion.

NIMBY is the bane of waste management.[21] Though everyone is happy to have their garbage collected and sewage flushed away, most want it to end up as far as possible from their 'backyards'. Solid waste disposal sites (dumps and landfills) are particularly unpopular, as much because of health and environmental concerns (landfill sites release 50–100 kilograms of methane per tonne of waste) as visual and olfactory offence.

Biogas production offers a way out, and the solution is as simple as it is effective: the landfill is sealed at the top and bottom, thereby creating a giant anaerobic digester, with pipes carrying the biogas straight from the landfill site to a power plant. Such an approach deftly circumvents nearly all the NIMBY arguments: waste is converted into a clean and valuable fuel, the impact on global warming is reduced, local jobs are created, and the plant may even become a beloved urban landmark (see Figure 4.45).

Anaerobic digestion has been used for decades, mostly in Europe and North America, to treat human and industrial waste. Thousands of waste treatment plants produce biogas for heat, power, and vehicle fuel. More recently, developing countries have been taking advantage of the United Nations' Clean Development Mechanism to earn carbon credits by capturing methane from landfill sites, such as the Bandeirantes Landfill in Brazil and the Nanjing Tianjingwa Landfill in China.[22] Several million low-technology digesters have

---

[20] Municipal solid waste, also known as urban solid waste, consists of matter discarded by humans. In countries without significant recycling systems, municipal solid waste predominantly includes food wastes, yard wastes, containers and product packaging, and other miscellaneous wastes from homes and businesses. Most definitions of municipal solid waste do not include industrial wastes, agricultural wastes, medical waste, or sewage sludge. In countries with a developed recycling culture, the waste stream consists mainly of intractable wastes such as plastic film and unrecyclable packaging.

[21] NIMBY is an acronym for the phrase 'not in my backyard', and describes opposition to a development project not on principle but because of its proximity to the objector.

[22] The Clean Development Mechanism is one of the procedures for emissions reduction defined in the Kyoto Protocol. It allows countries to gain 'credits' for steps

Figure 4.45. The egg-shaped digestion tanks of a municipal waste treatment and biogas plant in Hamburg, Germany, which have become a popular city landmark.
Source: Spirou42 on Flickr.

also been installed throughout Asia to convert manure, human waste, and farming residues into biogas for cooking and lighting, and digestate to fertilize crops (Al Seadi and Lukehurst 2012; IEA 2007; OECD/IEA 2008, 2009).

Converting Biogas to Biomethane

The energetic value of biogas is low compared with natural gas, as almost half of it comprises carbon dioxide, which doesn't burn. The carbon dioxide can, however, be filtered out, leaving biomethane, which can be used as a fuel in gas-powered vehicles or sold to a larger natural gas network. In some cases biomethane is already economically competitive with fossil fuels. In Austria and Germany, for example, many farmers grow crops for anaerobic digestion to produce fuel for their own tractors, and in Sweden more than 50 per cent of the fuel used for transport comprises biomethane (Fabbri and Soldano 2010; Petersson and Wellinger 2009).

> taken to reduce greenhouse gas emissions, such as capturing methane. These credits may then be traded with other countries or used domestically to increase the allowable quota of emissions.

Figure 4.46. A small biogas plant installed in Kerala, India, by BIOTECH, an NGO that has helped install biogas digesters in more than 20,000 households.
Source: www.biotech-india.org.

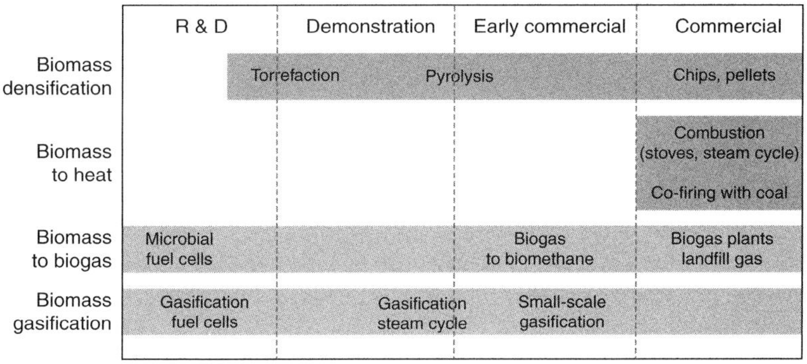

	R & D	Demonstration	Early commercial	Commercial
Biomass densification	Torrefaction	Pyrolysis		Chips, pellets
Biomass to heat				Combustion (stoves, steam cycle)
				Co-firing with coal
Biomass to biogas	Microbial fuel cells		Biogas to biomethane	Biogas plants landfill gas
Biomass gasification	Gasification fuel cells	Gasification steam cycle	Small-scale gasification	

Figure 4.47. Development status of the main bioenergy technologies.
Source: IEA Bioenergy (2009) (modified).

Gas from Solid Fuel

In 1609, Jan Baptista van Helmont, a Belgian chemist, discovered that by heating wood or coal, a highly flammable gas could be produced. Known today as syngas (or synthetic natural gas), this is a mixture of hydrogen, carbon monoxide, and methane. The production process for syngas uses heat, pressure and steam to break the molecular bond of any carbon-based material and convert it into smaller, more volatile molecules. This process was widely used in nineteenth-century England to convert coal into the 'town gas' that lit factories, homes, and streets. By replacing oil lamps and candles with a steady clear light, town gas played a major role in the Industrial Revolution, making night work possible in many industries. Town gas flickered out in the mid-twentieth century as it was replaced by electricity for lighting and by natural gas for cooking. However, gasification of solid biomass is still a mainstay of the chemical industry, where it is used to produce hydrogen, ammonia and other useful gases.

More recently, biomass gasification is being explored as a fuel source. Virtually any solid biomass feedstock (whether wood, straw or timber residues) can be converted into syngas, for use either as a fuel or as a raw material for a wide range of secondary conversion products. Since syngas consists of hydrogen, carbon and oxygen, the building blocks of all organic compounds, a recombination of these can produce valuable organic products such as methane, methanol, ethanol, diesel or plastics. Meanwhile, hundreds of small and medium-sized biomass gasifiers (< 1 megawatt) are being used, mainly for heating, in China, India, and Southeast Asia (IEA Bioenergy 2009).

Biofuels for Transport

"The use of vegetable oils for engine fuels may seem insignificant today. But such oils may become in course of time as important as petroleum and the coal tar products of the present time" (Lal and Sarma 2011, 15). With these words, spoken more than a century ago, Rudolf Diesel predicted that his engine, which was designed to burn less highly refined fuels, would one day run on biofuels. Diesel also put his money where his mouth was, presenting a prototype to the 1900 Paris Exposition that ran on peanut oil.

Today sugarcane is not only the mainstay of the modern food industry but also one of the most effective energy crops. The 135 kilograms of sucrose contained in one tonne of raw sugarcane can be transformed into 70 litres of ethanol, enough to fill the tank of an average car (da Rosa 2005). Brazil, the world's largest producer of sugar, was the first country to use ethanol as a motor fuel when the government introduced a 5 per cent blend with gasoline in the 1930s.

After the second oil crisis in 1979, when crude oil prices topped 40 U.S. dollars per barrel, Brazil turned to cane ethanol in a much more serious way. This process was aided by a huge spike in sugar production and a corresponding fall in prices. Numerous new distilleries were built, producing ethanol from both sugar and molasses and fuelling a boom in ethanol vehicles. By 1985 nearly all new cars sold in Brazil ran on ethanol only, though this trend was quickly reversed as the price of ethanol rose relative to gasoline. In recent years, the trend has been towards 'flexible fuel vehicles' (FFVs), which can run on gasoline, ethanol, or a mixture of the two. By 2007, FFVs accounted for 86 per cent of all new cars sold in Brazil and were becoming increasingly common in Europe and North America.

Brazil has stayed well ahead of other countries in its production and use of ethanol by developing a process in which sugarcane is processed into both ethanol and sugar, sidestepping the 'food or fuel' dilemma that has made maize-based biofuels so controversial. In 2006 Brazil was the world's undisputed sugar king, producing 33 million tonnes of sugar (more than 40 per cent of world demand) and 22.3 billion litres of ethanol (33 per cent of world production) (Soccol et al. 2010).

Despite Brazil's spectacular success in breaking, or at least reducing, its dependence on fossil fuels, biofuels have not caught on to a similar degree elsewhere. Though the production processes for biofuels are tried and tested, in 2008 they accounted for just 1.5 per cent of world transportation fuel. While that share is expected to grow, bioethanol and biodiesel are not about to displace petrol and diesel anytime soon.

The main reason for the modesty of biofuels' success is simple. For conventional (or first-generation) biofuels to seriously compete with fossil fuels, as they do in Brazil, vast tracts of agricultural land would need to be given over – fully or partially – to the cultivation of energy crops. On an already crowded planet, that is not a viable option. However, the biofuel sector is highly innovative. Already the technology to produce biogas from agricultural wastes (known as second-generation biofuels) is mature, though not yet exploited at a large scale, and there's a brave new world of new biofuels (third-generation) on the horizon.

First-Generation Biofuels

We tend to associate biotechnology with genetic engineering and clone-infested dystopias. Yet the ability to use plant and animal by-products to suit human needs goes right back to Noah's day – literally. According to the Bible (Genesis 9:20), Noah planted a vineyard on Mount Ararat, adding anecdote to the archaeological evidence that people living in what is now Armenia grew grape varieties suitable for wine production about 8,000 years ago (Vallee 1998).

Noah's contemporaries had not only mastered the art of selective cultivation, they had also learned to apply the natural reaction known as fermentation, one of the earliest forms of biotechnology. This involves microscopic unicellular fungi (yeasts) that feed on the sugars in fruit and vegetables releasing carbon dioxide and alcohol as waste by-products. Both of these by-products have been harnessed by humans; the former to make dough rise, the latter to make merry. After World War II, the use of fermentation spread from food production to the pharmaceutical and energy industries. Thus was born the first generation of biofuels; bioethanol from sugarcane or corn, biodiesel from vegetable oils and animal fats, and biomethane from the anaerobic digestion of wet biomass.

While grains and sugarcane store energy in the form of starch or sugars, other plants, such as olive, rapeseed, or palm, store their energy in the form of fats, in seeds or fruits. The process of exploiting plant oils is more mechanical than biochemical; the seeds are simply crushed or squeezed to release the oils. The vegetable oils we use in the kitchen are already suitable as fuels. They have a calorific value similar to diesel, just different physical properties. The main obstacle to the direct use of vegetable oils in cars is their density and viscosity.[23] Through a process known as transesterification, vegetable oils and animal fats can be converted into the more fluid and easily flammable biodiesel. Its energetic and physical properties are so similar to fossil diesel that a 5 per cent blend can be used in a normal diesel engine (Brittaine and Lutaladio 2010; IEA Bioenergy 2009).

Grassoline: The Promise of Second-Generation Biofuels

Most of the feedstocks used to produce first-generation biofuels are food crops. As more countries embrace biofuels, both as an alternative to oil and in response to climate change, their impact on the price of agricultural commodities grows. Already in 2007, a quarter of the American maize crop was destined for biofuel production (Kingsbury 2007). In 2008 there was a sharp rise in the global price of key food

---

[23] Vegetable oils are twenty times more viscous than diesel.

commodities, such as maize, wheat and soybean. Many, including the World Bank and several leaders of developing countries, were quick to blame biofuels, and the 'food versus fuel' debate made international headlines. This was often characterized as a choice between the gas tanks of the world's richest people and the bellies of its poorest. Jean Ziegler, the UN Special Rapporteur on the Right to Food, went as far as to call for a five-year moratorium on the production of biofuels. Subsequent analysis revealed that the main cause of high food prices was speculation by commodities traders (Baffes and Haniotis 2010). However, as long as global food and energy demands continue to increase, this debate is unlikely to abate. Both the Food and Agriculture Organisation and Greenpeace caution against too strong an embrace of biofuels for fear of destabilising food supplies (FAO 2011b; Graham-Rowe 2011; Harkki 2012).

Since all organic matter contains chemical energy, when we throw it away we are wasting energy. Every year 24 billion tonnes (100 EJ) of inedible plant material (such as cereal stalks and wood shavings), organic municipal solid waste, dung, and residues are produced as by-products of household, agricultural, forestry, and industrial activity (Arvizu et al. 2011b). While some of these by-products have traditionally been used, for example, as animal bedding and in construction, most are thrown away. Cellulose and lignin contain all the elements necessary to produce biofuels, and they are abundant. However, there is a snag: cellulose, unlike sucrose, is difficult to digest. But thanks to advances in industrial biochemistry over the past twenty years, we are now able to break down the complex molecules of cellulose or lignin (using heat and enzymes) into smaller, more digestible molecules, which can be further processed into a wide range of fuels such as bio-ethanol and biodiesel (Huber and Dale 2009; Sanderson 2011).[24]

Because the feedstocks are cheap, common, and don't compete with food supply, second-generation biofuels, or 'grassoline' as some witty scientists have dubbed them, are a very promising technology (IEA Bioenergy 2009). In 2011 work began on the world's first commercial-scale cellulose-based ethanol plant in Italy. This plant is expected to produce around 40,000 tons of ethanol a year from about ten times that weight of giant cane (*Arundo donax*), a bamboo-like grass that is traditionally used to make reeds for wind instruments such as saxophone or clarinet (Morales and Sulugiuc 2011).

---

[24] Enzymes that digest wood are used by the microbes that live in termite intestines or the wood-decomposing fungi that thrive on tree trunks.

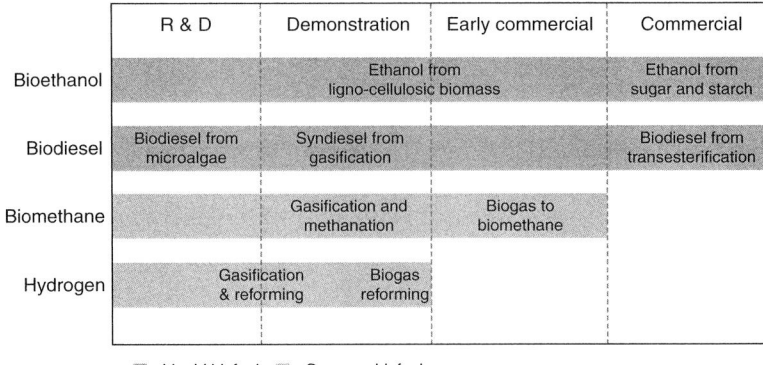

Figure 4.48. Development status of the main technologies to produce liquid and gaseous biofuels.
Source: IEA Bioenergy (2009) (modified).

However, second-generation biofuels are still largely unknown territory from a commercial perspective and represent an unpredictable and risky investment, exemplified by the case of Range Fuels, a company launched in November 2007 by former Apple executive Mitch Mandich. At first, this start-up attracted millions of dollars in private investment plus commitments for up to $156 million in grants and loans from the U.S. government. The plan was to build a biofuels plant that would convert wood chips and waste from Georgia's pulp and paper industry into ethanol. Within a year, the company had scaled back its production estimates from 40 million to 9 million gallons of ethanol a year. In 2011 Range Fuels closed its biorefinery without having sold a drop of ethanol (Biello 2011a).

Third-Generation Biofuels

When I was a university student, I would head out with my friends to one of the beach resorts along the Adriatic on summer weekends. I shied away from the discos because the volume cramped my style. I remember one full-moon night in particular, when my companion and I sat on the beach looking at the outline of our friends' bodies illuminated by sea sparkle (known colloquially as 'mare in amore' or 'sea in love' in Italy). Unfortunately, I got a little carried away explaining how the luminescent effect is caused by blooms of a unicellular organism and failed to take advantage of the romantic possibilities *Noctiluca scintillans* presented. I noticed two things that evening: unicellular organisms have a remarkable ability to rapidly reproduce, and there are times when it is better to look with the eyes of a poet than those of a scientist.

Algae like *Noctiluca scintillans* have the capacity to not only 'bloom', but also to produce and store oil in their cells. These twin characteristics make them the best hope for a third generation of biofuels. If cultivated in ponds under suitable conditions, algal microorganisms may yield up to 60,000 litres of biofuel per hectare, compared with 6,000 litres from a hectare of oil palm or 450 litres from soy (Demirbas 2010; Singh and Gu 2010). So far, third-generation biofuels are in the early stages of research and development, far from commercial production. But with the price of fossil oil and concerns about conventional biofuels on the increase, that sort of yield is already drawing considerable interest, both from scientists and investors (IEA Bioenergy 2009).

In 2010, the U.S. Department of Energy approved $44 million for research on algae-based biofuel. The interest from the private sector is even greater. In 2012 Sapphire Energy, a renewable energy company based in California, raised more than $150 million in private investment, bringing its total to more than $300 million. And the oil giant Exxon Mobil has invested $600 million in Synthetic Genomics, a company founded by Craig Venter, the American biologist-turned-businessman who led the race to map the human genome (Bullis 2012; Fairley 2011; Savage 2011). Venter summed up the advantages of investing in algae-based biofuels: "If we were trying to make liquid transportation fuels to replace all transportation fuels in the U.S., and you try and do that from corn, it would take a facility three times the size of the continental U.S. If you try to do it from algae, it's a facility roughly the size of the state of Maryland. One is doable, and the other's just absurd" (Biello 2012a).

Not content with his project to produce oil from algae, Venter is exploring an even more prodigious source of biofuel: bacteria that consume carbon dioxide and produce methane. Such organisms exist – deep in the ocean, surrounding thermal vents – but no one has yet succeeded in cultivating them in a laboratory, let alone in a power plant. However, with genetic engineering this may one day be possible. The great attraction of such fuel-producing bacteria is that they would solve two of our energy-related problems at a single stroke: removing the carbon dioxide that causes global warming, while supplying plentiful fuel to replace fossil fuels (Gibbs 2009).

Table 4.7. *Potential vegetable oil production in litres per hectare from different sources*

Species	Litres of vegetable oil per hectare
Soy	450
Peanuts	1,060
Rapeseed	1,200
Jatropha	1,900
Coconut	2,700
Oil palm	5,950
Unicellular algae	58,700

Source: Singh and Gu (2010).

Benefits and Costs of Bioenergy

Provided that the resources are developed and used sustainably and that efficient systems are used, bioenergy has the potential to reduce greenhouse gas (GHG) emissions by up to 90 per cent compared with fossil energy.

While solid biomass is already competitive with fossil fuels in household applications, biofuels are commercially competitive only in specific conditions, such as the highly productive environmental conditions of Brazil (abundance of agricultural land and a hot wet climate). Determining the production costs of energy from biomass is particularly complex because key factors vary between the many technologies available and locations. The main factors affecting the costs of bioenergy production are the cost of land and labour, crop yields, prices of inputs (such as fertilizer), water supply, transport distances from the land to the power plant, the scale of conversion (whether a domestic stove or a power plant), financing mechanisms and, last but not least, the price of fossil fuels. The estimated commercial bioenergy cost ranges are as follows:

- liquid and gaseous biofuels: 2–48 U.S. dollars (2005) per gigajoule;
- electricity or CHP systems more than 2 megawatts: 10–50 U.S. dollars per gigajoule (U.S. cents 3.5–25 per kilowatt-hour);
- domestic or district heating systems (solid waste to wood pellets): 2–77 U.S. dollars per gigajoule.

Recent analyses of lignocellulosic (second-generation) biofuels indicate potential improvements that will enable them to compete with oil prices of 60–70 U.S. dollars per barrel (0.38 to 0.44 U.S. dollars per litre). Depending on oil and carbon pricing, a combination of strong short-term research and development and market support could allow bioenergy to reach commercialization levels by around 2020 (Chum et al. 2011).

The Potential of Bioenergy

With 50 EJ of energy produced every year, biomass accounts for roughly 10 per cent of global primary energy supply. Global bioenergy use has steadily grown worldwide in absolute terms in the last forty years, with large differences among countries. In 2006 China was the world leader in the use of bioenergy (9 EJ), followed by India (6 EJ), the United States (2.3 EJ) and Brazil (2 EJ). The vast majority of this consumption is by the world's poorest people. Indeed, whereas biomass use in the industrialized world is associated with progress and technology, in the developing world it is still associated with subsistence.

In the largest developing countries (China, India, Mexico, Brazil, and South Africa), biomass accounts for up to one-quarter of primary energy production, and for as much as 80 per cent in parts of Africa. Although consumption in absolute terms continues to grow, the bioenergy share in India, China, and Mexico is decreasing, mostly as traditional biomass is substituted by kerosene and liquefied petroleum gas within large cities. By contrast, in many African countries, demand for wood fuels (mainly charcoal) has been increasing steadily in swelling urban areas.

In the highly industrialized countries, biomass accounts for 3 per cent of primary energy production (3.7 per cent if organic wastes are included). But here, too, especially in Europe, the share is growing through the use of modern biomass. The main sectors of bioenergy use are combustion of solid biomass and biogas for electricity generation, heat and steam production in paper plants and sugar factories, and combustion of pellets for residential heating (Chum et al. 2011; IEA 2012a; IEA Bioenergy 2009).

The current global crop (cereals, oil crops, sugar crops, roots, tubers, and pulses) is about 60 EJ/year, while global forestry production comes to 15–20 EJ/year (Chum et al. 2011). We can see straightaway from these figures that if bioenergy is to rise from its current 10 per cent and become a major source of future primary energy supply, the human appropriation of plant life on Earth will have to increase substantially. This would entail expanding agriculture and

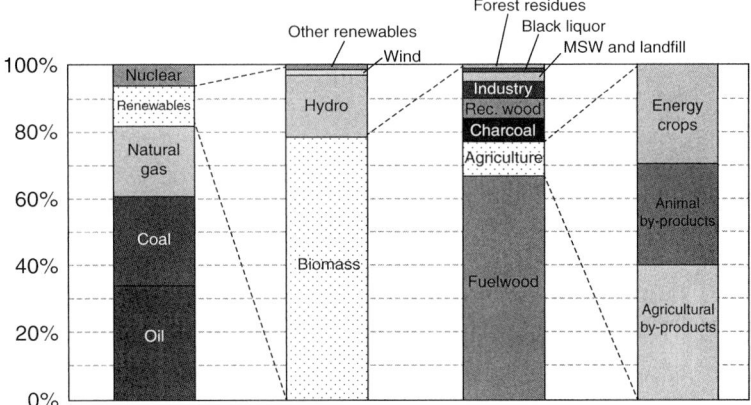

Figure 4.49. Breakdown of global bioenergy sources compared with the global primary energy mix. Renewable energy (RE) accounts for roughly 13 per cent of primary energy supply. Bioenergy is the most important of all renewables and represents roughly 10 per cent of total primary energy supply (almost 80 per cent of renewables). Of this, almost 80 per cent comes from forestry and agriculture, the rest is recovered wood, charcoal, wastes, and residues. See Table 4.8 for details.
Source: IEA website, IEA Bioenergy (2009), Chum et al. (2011).

forestry across a greater share of the world's landmass and increasing agricultural productivity through fertilizer, irrigation, and mechanisation. Meanwhile, world primary energy demand is expected to rise from its current 530 EJ to between 600 and 1,000 EJ by 2050 (IEA Bioenergy 2009).

The land requirements of the bioenergy economy are daunting. The annual global demand for liquid transportation fuel is equivalent to roughly two million tonnes of crude oil. Were all transportation based on biofuel crops grown in highly productive tropical regions such as Brazil, Indonesia, or Central Africa, 600 million hectares would be needed, more than the current total of tropical land under cultivation (Smil 2010, 100). Moreover, if we wanted to meet global primary energy production with sugarcane ethanol, we would need the entire global area currently under cultivation, and using biofuels from soy we would need the entire land surface of the Earth (Andrews et al. 2011; see Section 6.2 in Chapter 6). In short, we would need another planet: one for food production, the other for biofuels.

Assuming that efficiency increases in agricultural productivity keep pace with food demand, so that the share of land devoted to food

Table 4.8. Breakdown of global bioenergy sources compared with the global primary energy mix

Source	%	Renewables	%	Bioenergy	%	Agriculture	%
Oil	34.0	Bioenergy	78.8	Fuelwood	67	Agriculture by-products	40
Coal	27.0	Hydro	18.5	Agriculture	10	Animal by-products	30
Natural gas	20.6	Wind onshore	1.5	Charcoal	7	Energy crops	30
Renewables	12.7	Other renewables	1.1	Recovered wood	6		
Nuclear	5.7			Wood industry residues	5		
				Municipal solid waste and landfill gas	3		
				Black liquor (paper by-product)	1		
				Forest residues	1		

Sources: IEA website, IEA Bioenergy (2009), Chum et al. (2011).

production remains the same, the best chance of increasing the share of energy from biological sources rests with the ability to extend crop cultivation to less fertile lands that do not compete with food production and to exploit residues from forestry and wastes.

All in all, an optimistic estimate is that 100–300 EJ per year can be sustainably derived from bioenergy by 2050. It is reasonable to assume that biomass could sustainably satisfy no more than one-third of future global energy demand. However, there are large uncertainties in this potential, such as market and policy conditions, and it strongly depends on the rate of improvement in the production of food and fodder as well as wood and pulp products (Andrews et al. 2011; Chum et al. 2011; Eisentraut 2010; IEA Bioenergy 2009).

4.8 Geothermal Energy

Jules Verne was responsible, perhaps more than any other writer, for stimulating popular interest in science. His adventure stories have not only enthralled generations of children and adults; they also anticipated several major scientific innovations. Verne's tales of space and undersea travel may have prefigured scientific discovery, but his most popular novel, *Journey to the Center of the Earth*, was definitely more fantasy than science fiction.

Published in 1864, it describes how three men descend into a volcanic crater in Iceland, travel beneath the Earth's crust at a depth of more than 150 kilometres, and emerge through another volcano in Sicily. If someone were to try to replicate this adventure they would experience rapidly increasing temperatures in the outer crust (30°C per kilometre). At a depth of 150 kilometres, they would encounter sweltering temperatures of 600°C. Were they to make it to the core, as Verne's heroes intended, things would get very torrid indeed, with immense pressure and temperatures slightly higher than the surface of the sun (more than 6,000°C). While many of Verne's other fictional adventures, from deep-sea exploration to space travel, have been replicated, humankind's journey into the Earth has not gone very deep – so far, just 12.3 kilometres.[25]

Geologically speaking, the Earth consists of three layers: crust, mantle and core. The crust is the uppermost and thinnest

---

[25] In 2012 Exxon drilled the world's deepest well (12,376 metres) in the Chayvo oil field in eastern Russia. This is almost 2 kilometres deeper than BP's infamous Deepwater Horizon well that spilled 800 million litres of crude oil into the Gulf of Mexico in 2010.

layer – between 5 and 70 kilometres deep (it is thinner beneath the oceans and thicker beneath the continents). The mantle extends a further 2,900 kilometres down, while the core is a metal sphere, solid on the inside and liquid on the outside, 7,000 kilometres in diameter (roughly the size of Mars).

Some of the heat at the Earth's core has been trapped there by the crust blanket since the formation of the solar system, while the remainder derives from the radioactive decay of elements within the Earth.[26] But the Earth's layers are not homogeneous. In many areas of the world (Iceland, Hawaii, Japan, California, and East Africa to name a few), the heat from the core reaches the upper layers of crust, giving rise to volcanic activity and other geothermal phenomena. Geothermal occurrence involves hot volcanic rock heating wet sedimentary rocks that lie above. These rocks, in turn, heat groundwater, giving birth to phenomena such as hot springs and geysers (see Figure 4.52). There is also a steady supply of milder heat, sufficient to heat air or water to an ambient temperature, just about anywhere on Earth, starting at depths of just a few metres.

The word 'geothermal' derives from the Greek *geo* (earth) and *thermos* (heat), but its use by humankind predates Hellenic culture. Hot springs have been used for ritual or routine bathing since the Stone Age (Cataldi 1999) and perhaps for as long as our species has existed. In fact, we are not the only species to take advantage of hot springs. In the Jigokudani Monkey Park in Japan the resident snow monkeys (*Macaca fuscata*) while away many winter hours soaking in the warm waters of a hot spring, before returning to the forests at night (see Figure 4.50).

The oldest known spa was built in China during the Qin dynasty (third century BCE), and this practice spread throughout Europe under the Romans, who had a particular fondness for hot water.

The Devil's Power Plant

With steam hissing from the rocks, and a whiff of sulphur in the air, it is little wonder that a certain Italian valley came to be known as Valle del Diavolo. The ancient Romans, unperturbed by fear of demons, were the first to take a dip in the hot mineral-rich pools around the modern village of Larderello. Two millennia later, this became the site of the world's first geothermal electricity plant. An entrepreneur,

---

[26] Mainly thorium 232, uranium 238, and potassium 40.

Figure 4.50. Japanese macaque, better known as snow monkeys, taking a warm bath in a hot spring.
Source: Fg2 at Wikimedia Commons.

Prince Piero Ginori Conti, used the volcanic springs as a source of boric acid for chemical manufacturing, before hitting upon the idea of driving a turbine with the geothermal steam. For the next four decades Larderello was the world's only geothermal power plant. One hundred years later, Larderello is still steaming away, producing a tenth of the world's geothermal power and servicing about a million Italian homes.

Geothermal energy can be exploited at a range of temperatures, from lukewarm to steaming hot (approximately 30–300°C). It is easiest to harness in areas, such as Larderello, where magma (molten volcanic rock) pushes into the upper layers of the crust, allowing temperatures of more than 1,000 degrees Celsius to occur at depths of less than 10 kilometres. Such areas are suited to hydrothermal convective systems, where the magma heats groundwater circulating in permeable sedimentary rocks. The superheated water turns to steam and reaches the surface through natural fissures in the rocks, giving birth to geysers, fumaroles or hot springs. It is just a matter of capping the fissures and piping the hot steam to a nearby power plant. In the absence of natural fissures, wells can be drilled into the earth to retrieve the superheated water in much the same way as oil is extracted.

How Energy Is Produced 151

Figure 4.51. Since Roman times, steam and hot springs emerging from the geologically active ground of Larderello (Italy) has been used for hot water and bathing.
Source: woodcut by Guglielmo Jervis.

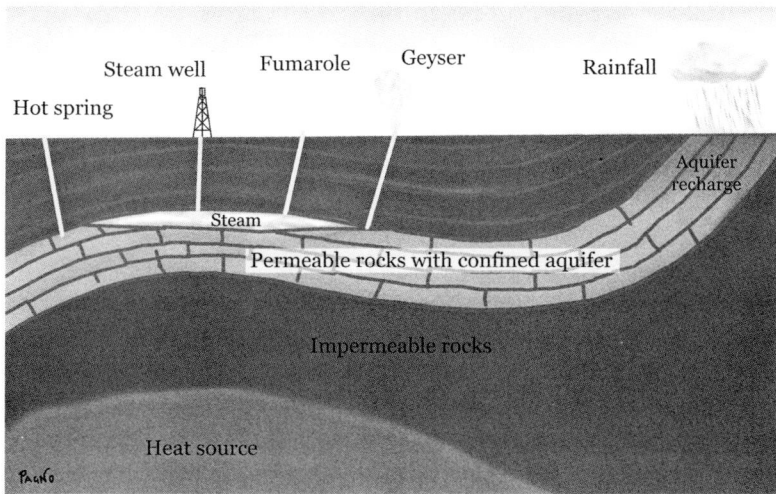

Figure 4.52. Characteristics of a hydrothermal site: an aquifer (e.g., permeable limestone) that receives a constant supply of rainwater is confined between impermeable rocks (e.g., clay sediments on top and granite at the bottom). The heat source (i.e., magma) forces hot water and steam upward through rock fissures, producing fumaroles, geysers, and hot springs.

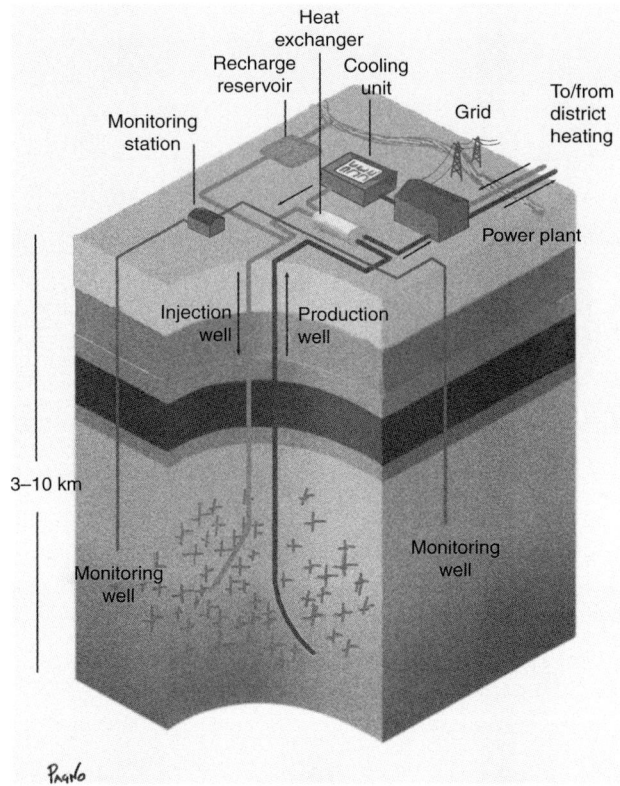

Figure 4.53. Diagram of an enhanced (or engineered) geothermal system.
Source: Goldstein et al. (2011) (modified).

Another promising, but not yet commercially viable, approach involves non-permeable dry rocks, where fractures are absent. In such cases hot water cannot be retrieved from an underground reservoir. Instead, holes are drilled into the rocks and pressurized water or a water-chemical mixture is injected, similarly to hydraulic fracturing (fracking) for shale oil and gas. The hot rocks act as a kind of giant underground hot plate around which water circulates (see Figure 4.53). This artificial circuit is known as an enhanced (or engineered) geothermal system (EGS).

The range of uses for geothermal energy is as varied as the methods of exploiting it. These include heat pumps in residential buildings, district heating, greenhouse heating, spas, factories, and power plants.

Figure 4.54. The Krafla Power Station is a 60-megawatt geothermal power station located near the Krafla Volcano in northern Iceland. Pipes collect steam and carry it to the power stations.
Source: Gian Andrea Pagnoni.

The possible applications vary depending on the geological environment, temperatures, and depths of the reservoirs.

Geothermal resources at high temperatures (more than 180°C) are used mainly for power production. So-called flash steam plants are the most common type of geothermal power plant. When hot water from the geothermal spring reaches the surface, it is captured in a low-pressure tank known as a separator. The emergence from very high pressure underground to very low pressure in the separator causes the hot water to quickly evaporate in so-called steam flashes. This blast of steam is drawn away to operate a turbine. Because the tank is much cooler than the underground reservoir, some of the steam condenses before reaching the turbine. This water, together with the condensed water that has passed through the turbines, is then pumped back into the geothermal reservoir.

Flash steam power plants work very well in places where the water or steam extracted is already at a very high temperature. Unfortunately, sites like this, where hot magma occurs close to the surface, are rare. However, some geothermal plants can generate power using water at temperatures as low as 60 degrees Celsius. These are known as binary cycle plants, because the geothermal energy is

Figure 4.55. The Sonoma Calpine 3 geothermal power plant at The Geysers field, 70 miles north of San Francisco in California, which was developed commercially in the 1960s. The steam powers twenty-two power plants with a net generating capacity of approximately 1 GW, enough to power the city of San Francisco.
Source: Stepheng3 at Wikimedia Commons.

exploited in two stages; the hot water extracted from the reservoir is first used to vaporize a secondary fluid, typically pressurized isobutane. Unlike water, which boils at 100 degrees Celsius, the secondary fluid vaporizes at a much lower temperature, and is therefore able to drive a (smaller) turbine. The binary principle allows geothermal power to be viable in more places, and this model is likely to predominate in the future.

Geothermal resources, particularly those at intermediate and low temperatures (between 60°C and 180°C) can also be used to supply direct heat. Heat exchangers or heat pumps transfer the heat from the hot water to a secondary fluid to heat residential buildings and greenhouses, or for industrial processes that require heat, such as food processing (IEA Geothermal 2012; Goldstein et al. 2011).

Ground-Source Heating and Cooling

There is something perverse about expending vast amounts of electricity to cool buildings, when there is an excellent source of cool air and water right beneath our feet. Ground-source heat pumps (GHPs), also known as geothermal heat pumps, exploit relatively constant underground temperatures and may be used to provide spatial heating, cooling, and domestic hot water for all types of buildings. GHPs do not rely on near-surface volcanic activity and can therefore be installed anywhere on Earth.

In temperate zones, where most of the world's energy is consumed, there is a constant temperature of around 15°C at a depth of ten metres, all year round. This stability may be exploited in winter to preheat water for conventional electric or gas-fueled boilers. If, for space-heating radiators, we need water at 55 degrees Celsius, but the water entering our home through municipal pipes is at 5 degrees, it makes sense to take advantage of the 15 degrees available just a few metres beneath our house. By raising the water temperature by just 10 degrees, we can shave 20 per cent off our gas or electricity bill. The way this works is remarkably simple. Water (or a secondary fluid) is injected into pipes that follow a loop to a depth of 5–10 metres. The fluid returning from the loop is conveyed to a heat pump, an electrical vapour compression unit that works like a fridge when cooling or a reversed fridge when heating. In summer the underground stable temperature will be used for cooling applications. In terms of geology, GHP systems bear little relation to the recovery of volcanic heat at sites like Larderello. Strictly speaking, GHPs do not exploit geothermal energy, but rather solar energy stored under several metres of insulating earth.[27]

There are two main types of GHP system: closed-loop and open-loop. In closed-loop systems, a circuit of plastic piping is placed in the ground, either horizontally at a depth of 1–2 metres or vertically at 50–250 metres. A water-antifreeze solution is circulated through the pipe, gaining heat in winter and losing it in summer. An open-loop system pumps groundwater or lake water directly from its source and then discharges it, either into another well or back into the same water reservoir (Lund 2007; Rybach 2005).

The Benefits and Costs of Geothermal Energy

Geothermal energy is classified as a renewable resource because the tapped geothermal heat is naturally replenished by the ongoing nuclear decay in the Earth's core. Within a human timescale, this energy is inexhaustible. Geothermal energy has the potential to provide long-

---

[27] GHP efficiency is theoretically high as with all electrical devices (up to 90 per cent), but real efficiency must take into account that electricity is normally bought from the grid, where electricity is produced through fossil fuel-based conventional plants that have an efficiency of 30–40 per cent.

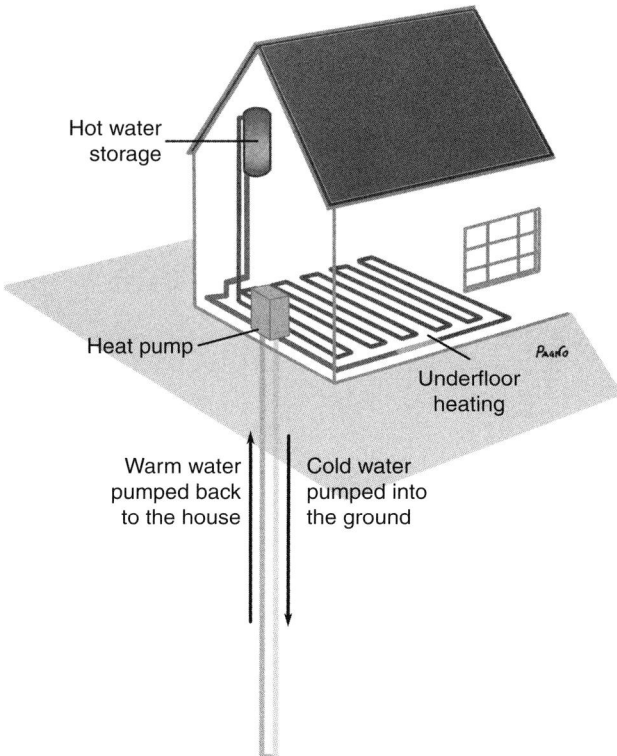

Figure 4.56. A GHP circulates a mixture of water and antifreeze around a loop of pipe (the ground loop) beneath or next to the building. In winter, heat from the ground is absorbed into this fluid and is pumped through a heat exchanger (vice versa in summer). The length of the ground loop depends on the size of your home and the amount of heat you need – longer loops can draw more heat from the ground but need more space to be buried in.

term, secure base-load energy and reductions in greenhouse gas emissions (Goldstein et al. 2011).

Geothermal projects typically have low operational costs but high initial investment costs because of the need to drill wells and construct power plants. Cost estimates for geothermal power installations vary widely between countries, from 3.1 to 13 U.S. cents (2005) per kilowatt-hour for flash steam plants and 3.3–17 U.S. cents per kilowatt-hour for binary cycle plants. EGS projects are expected to be more capital intensive, however, since there are no commercial EGS plants yet in operation (Goldstein et al. 2011).

The Potential of Geothermal Energy

In 1950, geothermal power generation was confined to two countries: Italy and New Zealand. Today, twenty-four countries generate electricity from geothermal resources and global installed capacity has risen fiftyfold, from 0.2 to 11 GW (see Table 4.9). Yet, though encouraging, this increase is frightfully modest compared with the spectacular growth of other renewable energy resources, particularly solar PV and wind. The reason geothermal power remains marginal is that, although high-grade geothermal resources are already economically competitive with market energy prices in many locations, they have restricted geographic distribution, while the use of low-grade geothermal resources and EGS encounters both cost and technology barriers. That is why commercial exploitation to date has focused on shallow (less than 400 metres) reservoirs with high fluid temperatures (more than 180°C).

The world's top five geothermal power producers are the United States, Philippines, Indonesia, Mexico and Italy. Since crust heat is by definition widespread, this resource is receiving increasing interest, and many countries increased their capacity by more than 50 per cent between 2005 and 2010 (REN21 2012).[28]

In six countries geothermal energy satisfies more than 10 per cent of the electricity demand, and in seventy-eight countries it is used directly for heating and cooling. The worldwide use of geothermal energy for power generation is roughly 0.25 exajoules per year (70 TWh per year), whereas GHP applications generate roughly 0.44 exajoules per year (121.7 TWh per year) of thermal energy.

The current global capacity of geothermal electricity corresponds to fewer than ten conventional coal power plants. The Earth's potential to donate warmth (5.4 billion EJ per year in the Earth's crust alone) is huge if compared with current global primary energy supply (530 EJ per year). Clearly, we are sitting on a vast potential resource, yet the estimated technical potential varies depending on technology development and the depth at which the resource is expected to be exploited.

For electricity generation, assuming drilling to a depth of 3 kilometres, the technical potential is estimated at 118 exajoules per year, one-fifth our current global energy production. But 3 kilometres is still

---

[28] Notably Germany, Papua New Guinea, Australia, Turkey, Iceland, Portugal, New Zealand, Guatemala, Kenya, and Indonesia (Bertani 2012; REN21 2012).

Table 4.9. *Installed capacity for geothermal electricity generation in 2010 in the main producer countries*

Country	MW
United States	3,098
Philippines	1,904
Indonesia	1,197
Mexico	958
Italy	843
New Zealand	762
Iceland	575
Japan	535
El Salvador	204

Source: Bertani (2012).

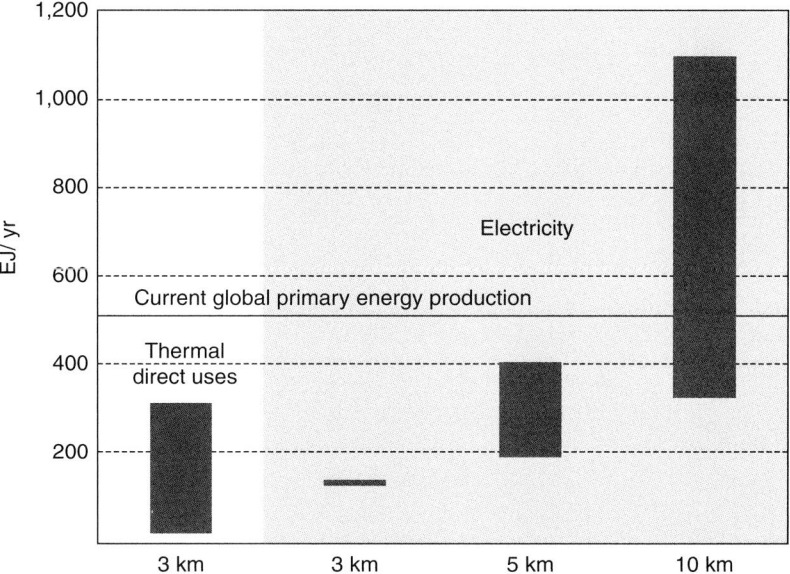

Figure 4.57. Ranges of technical potentials of geothermal energy for direct (heat) uses at a depth of 3 kilometres, and for electricity generation (at 3-, 5-, and 10-kilometre depths).
Source: Goldstein et al. (2011) (modified).

relatively shallow. Geothermal wells are currently drilled to depths of 5 kilometres using methods similar to those used for oil and gas exploration. If advances in technology enable drilling down to 10 kilometres, geo-electricity potential could rise to 1,100 exajoules per year, more than twice global primary energy production. Most of this potential is ascribed to EGS (Biello 2013b), while the technical potential of direct thermal uses is estimated to range from 10–312 exajoules per year (Dickson and Fanelli 2004; Goldstein et al. 2011).

4.9 Ocean Energy

Between two-thirds and three-quarters of our planet's surface is covered by contiguous seas and oceans. This huge mass of salt water has always played a central role in human development, acting both as a conduit and a constraint. The Mediterranean region became the cradle of Western civilization precisely because that body of water facilitated exchange between the peoples that settled on its coasts. By contrast, the oceans were for millennia the great unknown, a place of unfathomable depths and terrible monsters, the inspiration for innumerable sailors' yarns. Even in the modern age, when satellite images map out almost every inch of landmass, the oceans are the least known parts of the world.

In terms of energy, oceans represent an immense and largely untapped resource. For centuries, basic mechanical techniques to generate energy from the tides and currents have been known, but these have only been pursued consistently since the 1970s. Ocean power languished in the post-oil-crisis period of the 1980s, but at the turn of the millennium, amid growing concern about peak oil and climate change, research and development of a wide range of ocean energy technologies gained a new lease of life (JRC 2011; Lewis et al. 2011).

King Canute's Lesson

King Canute famously demonstrated that the oceans do not respond to human command.[29] However, human ingenuity has managed, if not to

---

[29] King Canute, or Cnut the Great, was king of England from 1016 to 1035. Legend has it that Canute commanded the tide to turn back. Chroniclers have interpreted this as an act of either foolish arrogance or of piety (to demonstrate the limits of worldly power).

tame, then at least to harness the great power of the oceans. Several different methods have been developed. In summary, these are the following:

- Wave power: The rhythmic motion of the ocean surface is transformed into electricity.
- Tides and currents: Water flow, driven by tides or ocean currents, passes through a barrage containing water turbines.
- Ocean thermal energy conversion (OTEC): The difference in temperature between the surface and the depth of the oceans is exploited either for heating or to operate a turbine.
- Salinity gradients: at the mouth of rivers, where freshwater mixes with saltwater, differences in salinity can be harnessed to produce electricity (Lewis et al. 2011).

Wave Power

As the wind blows over the ocean, some of the wind's energy is transferred to the water, forming waves. The rise and fall of waves is a form of kinetic energy that may be harnessed similarly to the piston in an internal combustion engine. This linear movement (up and down) can be converted into a circular movement of a shaft, which is, in turn, connected to a generator (see Figure 4.58) (Lewis et al. 2011). But this is not the only way. More than fifty different models of wave energy plant, either floating or anchored to the sea bed, are currently under development.

Tides and Currents

Ocean waves, moved by winds, are a form of indirect solar energy. Yet, when we talk about ocean energy another body of the solar system comes into play: the moon. The gravity of our satellite is one-sixth that of the Earth, enough to cause slight fluctuation in the Earth's orbit and to attract the Earth's waters (see Figure 4.59).[30]

---

[30] As the moon orbits the Earth, it draws water away from the Earth's surface. The water masses facing the moon are attracted, causing high tide, while on the opposite side of the Earth, water is drawn away from the surface by centrifugal force (of the Earth-Moon system), causing what are known as tidal bulges. Low tides occur because, as the waters are drawn into the two bulges they must recede at the points least affected by the moon's gravity.

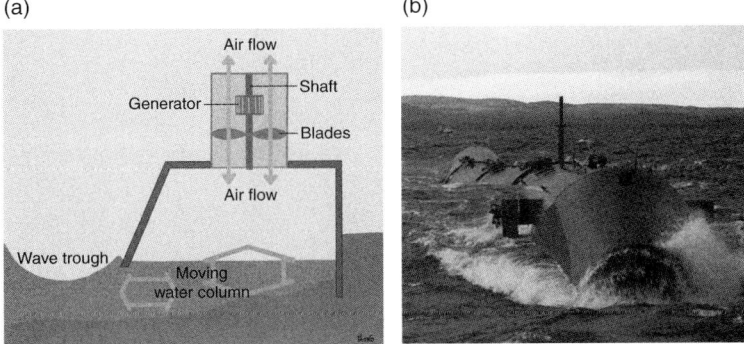

Figure 4.58. Two types of wave energy converters. In the model on the left, waves entering and exiting a chamber causes the air in the chamber to rise and fall, which, in turn, drives a turbine. The second device comprises five tube sections linked by flexible joints. The movement of these joints caused by waves is converted into electricity via hydraulic systems housed inside each joint. Power is transmitted to shore using subsea cables.
Source: Lewis et al. (2011) (modified) and www.pelamiswave.com.

Seas and oceans experience two high tides and two low tides per day. The tidal range depends on the dimension and depth of the water bodies. Enclosed seas, such as the Black Sea or the Baltic, experience little or no tidal movement, while the oceans experience tides up to several metres. At the top end of the scale is Wolfville, in Nova Scotia (Canada), where tethered boats rise and fall 16 metres (the highest tidal difference on Earth) every day.

Up to now, tidal energy has mostly been harnessed on estuaries closed off by a barrage (see Figure 4.60). The principle at work is much the same as in a hydroelectric plant (within the barrage are water turbines that generate electricity), except that the water basin is replenished not through rainfall but through tidal flow.

The biggest problem with such plants is that they interfere with the movement of ships and marine animals and disturb delicate estuarine environments. However, recent advances allow tidal plants to be located far from estuaries, offering greater flexibility in terms of capacity and impact.

Under certain conditions the kinetic energy of tidal or ocean currents may be harnessed without a barrage using turbines that exploit tidal or ocean currents. The difference between these is that tidal currents change direction between ebb and flow cycles, while ocean

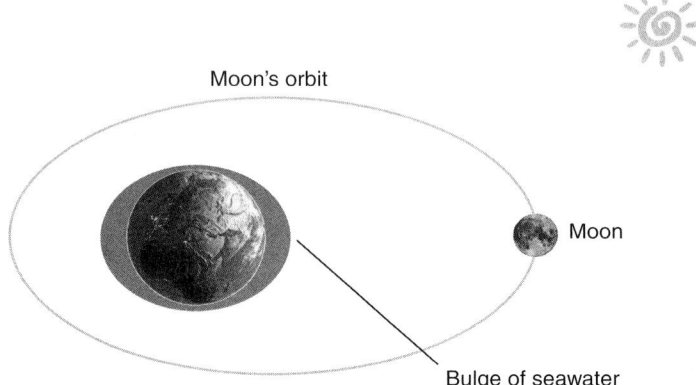

Figure 4.59. How the moon influences the tides.

currents have a unidirectional flow. Tidal and ocean current turbines are very similar; the main difference being that tidal installations have been designed to operate in both directions. The turbine has a horizontal axis and more closely resembles turbines than to hydro turbines (see Figure 4.61) (Lewis et al. 2011).

Ocean Thermal Energy Conversion (OTEC)

About 15 per cent of the solar thermal energy absorbed by the oceans is retained in surface waters. In tropical latitudes, average sea surface temperature can even exceed 25 degrees Celsius. Twenty-five degrees are obviously not sufficient to drive a steam turbine, but this warm seawater from the ocean surface may be used to vaporize a fluid with a low boiling temperature (such as ammonia, propane, or chlorofluorocarbon). This vaporized fluid drives a turbine, is then cooled by seawater recovered from the deep sea, condenses as liquid, and re-enters the cycle. Ocean energy can also be used to provide air-conditioning, district heating and cooling in coastal locations in a way similar to geothermal energy (Lewis et al. 2011).

Salinity Gradients and Osmotic Power

The reason it is dangerous to use a hair dryer or other electrical appliance in the bath is not because water conducts electricity (it doesn't), but because the salts present in freshwater do. In a battery or between the poles of a socket, electrons flow from an area of high electron

How Energy Is Produced   163

Figure 4.60. Aerial view of the tidal barrage and 240-megawatt tidal power station at the estuary of the River Rance at Saint Malo, France.
Source: Tswgb at Wikimedia Commons.

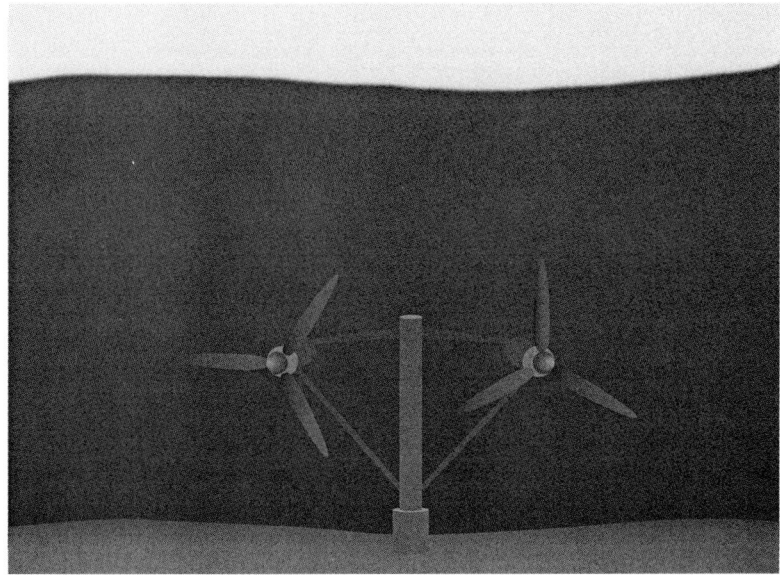

Figure 4.61. Horizontal tidal current converters. The turbine blades are a hybrid of wind turbine and ship's propeller.
Source: Lewis et al. (2011).

density to an area of low electron density. Something similar happens if we add salt to a pot of boiling water. As the grains of salt dissolve, the molecules of sodium chloride split into ions (molecules with an electric charge); in this case, positive sodium ions and negative chloride ions.

When a river flows into the sea, freshwater and saltwater mix. So-called reversed electro dialysis (RED) systems can exploit the difference in electric potential between two solutions with different concentrations of ions. Saltwater and freshwater are brought into contact through an exchange membrane,[31] and the difference of charges that results may be used to generate a current. The first prototype to test this concept is currently being built in the Netherlands.

Another way to take advantage of different concentrations of salt in water is osmotic power. In this case, the osmotic potential rather than the electric potential of ions dissolved into the seawater is harnessed and a different type of exchange membrane is used. A semipermeable membrane allows water molecules to pass through but prevents salt molecules from doing the same. For physical and chemical reasons, the concentration levels at the two sides of the membrane tends to balance out and water molecules naturally move through the membrane from the region of low to high concentration of salts (from freshwater to salt water). The osmotic pressure difference may be up to 26 bar, roughly ten times the pressure in a car tire. This process produces a small but consistent stream of water moving from one side of the membrane to the other. This flow can then be harnessed to drive a hydropower turbine. The world's first 5-kilowatt osmosis power plant was commissioned in Norway in 2009 (Lewis et al. 2011).

The Benefits and Costs of Ocean Energy

Although more than 100 different technologies are under development in more than thirty countries, ocean energy systems are at a very early stage and will not have much impact on energy markets for many years. Some governments, particularly in countries along the northeastern Atlantic, are promoting ocean energy technologies.

---

[31] An ion exchange membrane is a semipermeable membrane designed to conduct ions while being impermeable to other substances such as water or gases. In normal conditions, positive and negative ions would mix into the solution, whereas this semipermeable membrane allows the separation of positive and negative ions. A concentration of positive and negative charges occurs at different sides of the membrane, making the system like the two poles of a battery.

With the exception of tidal power, which basically adapts the technology of hydroelectric dams to estuarine situations, most ocean energy technologies have not yet progressed beyond the prototype stage.[32] Even tidal power is in its infancy. A small conventional coal plant generates more electricity than the world's total installed capacity of tidal power of roughly 600 megawatts (end of 2011). However, many projects are in development all over the world, some of them with very large capacities, including in the UK (Severn Estuary), India, Korea and Russia. The total capacity under consideration is 43 gigawatts (64 TWh per year), the equivalent of roughly thirty conventional coal-fired power plants.

The Potential of Ocean Energy

Until about 2008, ocean energy was not considered in any of the major energy models, and therefore its potential impact on future world energy production is just beginning to be investigated. Some ocean energy resources, such as ocean current converters and osmotic power, can be used anywhere. Others are more localized: OTEC is mostly available in the tropics, while the best waves occur at medium latitudes (30 to 60 degrees). The global theoretical potential for ocean energy technologies is estimated at 7,400 exajoules per year, roughly fourteen times the global primary energy production (530 EJ per year). However, there is substantial disagreement about the technical potential, which is estimated between 7 and 331 exajoules per year. Since the upper estimate proposes 'ocean thermal' as the energy with highest potential (300 EJ per year) followed by 'wave energy' (20 EJ per year), perhaps in the future, heat and cooling will be the main form of energy harvested from the oceans (Lewis et al. 2011).

4.10 New Frontiers: Science or Fiction?

The Power of the Stars: Nuclear Fusion

When, in 1905, Albert Einstein, unveiled his famous formula ($E = mc^2$), proposing that mass is equivalent to energy, scientists still believed

---

[32] Of the several technologies under development, the only commercially operational ocean energy technology is the tidal barrage, of which the best examples are the 240-megawatt La Rance Barrage (France) and the 254-megawatt Sihwa Barrage (South Korea).

that the sun was a sphere of molten rock. A few decades later, spectrographic studies showed that the sun is composed mostly of hydrogen, with some helium. In 1938, the physicist Hans Bethe revealed why: the pressure at the core of our star is so great that it fuses hydrogen atoms to produce helium. Every kilogram of hydrogen is converted into 993 grams of helium. The 'lost' 7 grams of mass is released as energy; 63 trillion joules for every kilogram of hydrogen.[33] If this could be replicated in a reactor, less than 1,000 tons of hydrogen would be needed to meet the entire current energy needs of the human race. This is very little compared with the 4.3 billion tonnes of oil extracted every year. It gets even better: hydrogen is so common on Earth that reserves could outlast the lifetime of the solar system, and helium, the product of hydrogen fusion, is completely non-toxic. No wonder hydrogen fusion is the energy dream that has kept generations of physicists awake at night.

Scientists have managed to reproduce hydrogen fusion on Earth, but not in a reactor. In 1952, the first hydrogen bomb was exploded on the Enewetak Atoll in the Pacific Ocean, releasing 450 times the energy of the fission-based bomb that was unleashed over Nagasaki seven years earlier. Releasing such a massive burst of energy is one thing; harnessing it to produce electrical power is another.

The greatest challenge with hydrogen fusion is getting the reaction started. Because the nuclei of hydrogen atoms are positively charged, they repel one another like the equivalent poles of magnets. To bring about fusion, it is therefore necessary to slam hydrogen atoms into one another with great force. The hydrogen bomb gains this initial burst of energy from a small fission-based atomic bomb within the missile shell. Thus, fission ignites fusion. Temperatures of 100 million degrees Celsius would be needed to initiate an industrial-scale fusion reaction. No known material can withstand such temperatures, and without a container the reaction cannot be controlled.

This limitation has not prevented scientists from experimenting with fusion in the laboratory. Two main fusion experiments are currently underway: the National Ignition Facility (NIF) in the United States and the International Thermonuclear Experimental Reactor (ITER) in France. The first uses 190 concentrated laser beams to heat

---

[33] According to Einstein's formula $E = mc^2$, $E = 0.007 \text{ kg} \times (300{,}000{,}000\text{m})^2 = 6.3 \times 10^{14}$ J.

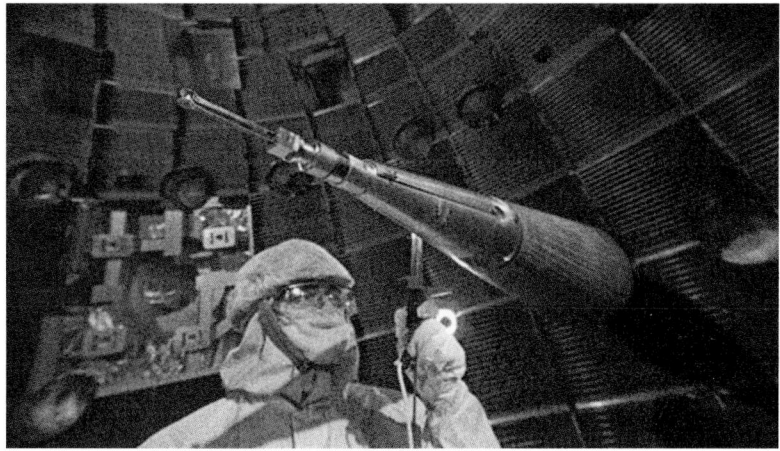

Figure 4.62. A technician inside NIF target chamber.
Source: Lseaveratnif at Wikimedia Commons.

a small amount of hydrogen fuel to the point where the atoms fuse. The second is a circular magnetic chamber that compresses atoms to achieve fusion and prevent the reactor wall from melting.[34]

Experiments around the NIF idea began in the late 1970s, and following a series of cost overruns, the US$3.5 billion NIF project was launched in 2009. Similarly, the ITER project was launched in 1985, construction started in 2008, and the end of the project is planned for 2038. The initial goal of ITER sounds modest: to keep hydrogen plasma burning for just a few seconds, producing an energy output ten times greater than the energy input. Even if this goal is achieved, it will require a major additional leap to develop a commercial reactor. As with most technological innovations, the issue of cost is paramount. Already, the ITER budget has tripled to around US$20 billion. Even with unlimited funding, it would take between thirty and seventy-five years before a large-scale fusion reactor can be constructed.

Much will depend on how other alternative energy technologies fare in the meantime. A breakthrough in solar or other renewables may put the brake on expensive fusion research (Brumfiel 2012; Gibbs 2009; Moyer 2010; www.iter.org).

---

[34] The ITER project is funded and run by seven member entities: the European Union, India, Japan, China, Russia, South Korea, and the United States.

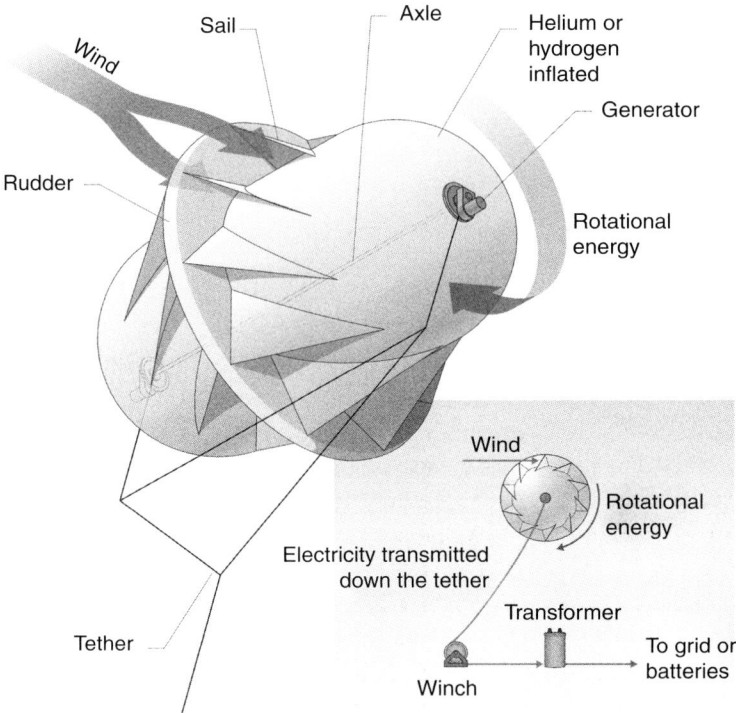

Figure 4.63. Model of a helium-inflated airborne wind turbine.
Source: James Provost at Wikimedia Commons (modified).

Windmills in the Sky

Two-thirds of the wind movement on this planet occurs in the upper troposphere (the lowest level of the Earth's atmosphere), where incessant wind currents, known as jet streams, swirl around the globe. This, in wind energy terms, is the mother lode. Since it is not possible to construct towers that rise 10,000 metres into the sky (the world's tallest building is 'just' 830 metres tall), engineers are looking at other ways to position turbines at such heights. These typically involve various types of kites and balloons that are tethered to the ground.

The great attraction of airborne wind turbines is that they would transform wind power from an intermittent to a base-load power source. Moreover, they would dispense with the most expensive component of conventional wind power plants – the tower – and greatly reduce the impacts in terms of land use and bird mortality. However, as good as this sounds, there are a few major technical questions that need to be answered before airborne wind turbines progress from the experimental to the commercial stage: how to safely suspend devices thousands of metres above the ground, how high the maintenance costs will be, and how to prevent collisions with aircraft (Gibbs 2009).

Cold Water on Cold Fusion

In 1989, Stanley Pons and Martin Fleischmann made a splash. Their claim to have initiated a fusion reaction at room temperature unleashed a frenzy of scientific and public interest. The year 1989 was already one of extraordinary events. Starting in March, a few weeks before Pons and Fleischmann's announcement, the Iron Curtain began to tear. The world's media was highly receptive to stories of breakthrough, and the prospect of unlimited safe and clean energy that this claim opened up was more than revolutionary; this was the modern equivalent of the philosopher's stone, the legendary substance, eagerly sought by medieval alchemists, that could turn base metals into gold. Unfortunately, unlike the events unfolding in Eastern Europe, Pons and Fleischmann's news turned out to be too good to be true. Under pressure from the university that had funded their research, they had made their announcement prematurely, before submitting their findings to peer review. After numerous scientists tried, and failed, to replicate their experiment, the claim was discredited.

Since 1989, numerous scientists have tried to demonstrate cold fusion, but all have failed. This quest may turn out to be as illusory as the search for the philosopher's stone, but at least it has stirred public interest in the energy debate, particularly in the potential of hydrogen fuel (Gibbs 2009; Saeta 1999).

The Enterprise Solution: Matter-Antimatter Reactor

When, in the television series *Star Trek*, Captain Kirk boldly embarked on the next adventure, he relied on his spaceship's antimatter engine. Though pure fiction, there was some science behind this. Antimatter is almost identical to ordinary matter, except that its subatomic particles are charged oppositely to normal particles; protons are negative and electrons are positive. Normally, atoms repel one another because of their similar charges. That is why immense heat and pressure are required for atomic fusion. With atoms of antimatter, the opposite occurs; they attract one another with a force equivalent to the repulsion of normal atoms. As a result, when they collide their mass disappears and they are converted into energy. Antimatter exists naturally in the universe, but it is very rare. It can be synthesized in particle accelerators, but even the world's most efficient accelerator – at CERN in Switzerland – would have to run nonstop for 100 billion years to produce a single gram of antiprotons (Collins 2005; Gibbs 2009). However, there is an even bigger problem with antimatter: any tool (composed of matter) used to handle or contain antimatter would be immediately annihilated by it.

Space-Based Solar Energy

In 1968, Peter Glaser, then president of the International Solar Energy Society, proposed an energy solution for the space age: giant satellites orbiting the Earth would capture solar power and beam it to the ground in the form of microwaves or laser. During the 1970s oil crises his idea garnered considerable interest, but when cost estimates of about US$300 billion were revealed, the interest quickly evaporated.

However, the idea, in itself, is not outlandish. Sunshine is eight times more intense in the Earth's upper atmosphere than it is on the ground. From 1995 to 2003 NASA funded several projects that evaluated a variety of solutions. Designs took advantage of thin-film PV panels to generate electricity, which could then be converted on board the satellite into electromagnetic waves and transmitted at low frequency to ground stations. Here, receivers 2 to 3 kilometres wide would convert the waves back into electricity. Like visible light or radio signals, microwave radiation does not damage DNA and biomolecules and is therefore safe in low doses.

The biggest obstacle to space-based solar energy is not technology but economics. The food that astronauts eat on the International Space Station is almost worth its weight in gold, as it costs more than US$10,000 per kilogram to launch goods into space. Unless these costs can be reduced to a few hundred dollars per kilogram, space-based solar technology will not be competitive with terrestrial plants and will only be used by astronauts (Gibbs 2009; Hadhazy 2009).

5
Challenging Times
The Politics and Economics of Energy

5.1 The World Markets in Oil, Gas and Coal

At the end of World War II, seven Western oil companies controlled most of the world's oil production,[1] and only the national oil companies of the USSR and Mexico could rival them in terms of production. These so-called 'majors' worked the bountiful oil fields that straddled the Persian Gulf, and their dominance helped to fuel America's economic expansion and Europe's recovery after the war.

The first challenge to this dominance came in 1950 when the king of Saudi Arabia negotiated an increased share in oil revenues with the U.S. majors. The following year, the new Iranian prime minister, Mohammad Mossadegh, attempted to nationalize his country's oil. Before Mossadegh could do so, the British and American secret services engineered his overthrow in a coup d'état (Fisk 2006). In 1956, Britain again intervened to safeguard oil supply, this time joined by France and Israel, following Egypt's nationalization of the Suez Canal, the main route for tankers sailing from the Persian Gulf to Europe. In the end, the Suez Crisis weakened Europe's influence in the Middle East and hardened the resolve of petroleum-producing states to control their own resources. Gradually, over the next twenty years, they did so, most significantly through the establishment in 1960 of the Organization of Petroleum Exporting Countries (OPEC).[2] By 1970, with oil production nationalized in most member countries, OPEC was in a position to control the global price of oil.

---

[1] These were Chevron, Texaco, Exxon, Gulf, Mobil, Anglo-Iranian (now BP) and Shell.
[2] OPEC was founded by Iraq, Iran, Kuwait, Saudi Arabia, and Venezuela; it expanded over the next decade to include all the main Middle Eastern and African producers.

In 1973 OPEC became a household term in the industrialized world. The Arab members of the cartel imposed an embargo on oil sales to protest against American support for Israel in the Yom Kippur War. The effect was a quadrupling of the price of oil, which sent a shock wave through the Western economies. One of the responses of the West was to establish the International Energy Agency (IEA), an organization representing the major oil consumers, which, it was hoped, could counteract OPEC. Members are required to maintain stocks of oil equivalent to ninety days of the previous year's consumption. In the event of cut-offs or restrictions to the flow of oil, IEA members can release oil from their stocks, thus stabilizing the market.[3] This would obviously help to mitigate an emergency situation but does not offer a long-term solution. More recently, the IEA has focused on advising members on energy policy and ways to move away from fossil fuels towards more sustainable energy sources.

Because oil production is dominated by OPEC and the 'majors', the world oil market is anything but free. OPEC members, who control about 60 per cent of current production, can manipulate the price of a barrel of oil by increasing or reducing their own production. It is a crude mechanism, and one that has threatened to backfire on several occasions (high oil prices provide an incentive for new exploration and for recourse to alternatives), yet more than fifty years after its creation, OPEC remains the biggest player in the global oil market (Yergin 2011).

One of the most significant responses of oil-consuming states to the insecurities of the market has been to embrace natural gas. Because of the technical and safety challenges involved in extracting, transporting and using natural gas, it took longer for a gas market to develop. Initially, gas was a national or regional fuel. Only with the development in the mid-twentieth century of long-distance pipelines and the technology to produce liquefied natural gas (LNG) did a global market emerge. The natural gas market is similar to the oil market. Indeed, many of the same players are involved, as natural gas is generally found in combination with petroleum. However, the gas market is a lot more stable. This is because the high costs associated with making and moving LNG have induced the various parties – countries, oil companies, power utilities and trading houses – to sign long-term

[3] This mechanism has only been used three times since the organization's foundation in 1974; during the Gulf crisis of 1990–1991, in the aftermath of hurricanes Katrina and Rita in 2005, and during the civil war in Libya in 2011.

contracts. Furthermore, thanks to increased competition between Middle Eastern and Russian gas suppliers, and the recent shale gas boom in North America, the global gas supply has both increased and become more diverse (Yergin 2011).

There has been a lot of talk in recent years about a renaissance of coal, but the truth is that coal never really went away. It was replaced by oil as a transportation fuel in the early twentieth century but remained the fuel of choice for electricity generation throughout the century. If the world coal market is booming today, it is not because of any great change in the energy mix, but simply because the demand for electricity, particularly driven by China and India, is growing rapidly. Coal is available in large quantities in most of the industrial countries, so the market is stable and predictable (Verein der Kohlenimporteure e.V. 2010; Yergin 2011).

5.2 Free and Not-So-Free Energy Markets

Major disruptions in the energy economy – from a spike in the price of oil to an accident at a nuclear power plant – send shock waves through the global economy. The energy sector is unique in this way; no other is as indispensable to all others, and few are as lucrative. Energy companies are also among the largest and most powerful corporations in the world.

Because many activities in the conventional energy sector, from exploration and mining to the construction of power plants, are so expensive, only large players tend to survive. In 2012, six of the world's ten most profitable companies (by revenue) operated in the energy sector.[4] It is hardly surprising, then, that these companies are able to exert enormous influence on governments. A market in which a few large players dominate, in which the largest players are state-run monopolies, in which governments intervene – diplomatically, economically, and militarily – to safeguard supply, and in which the largest producers are autocratic regimes can hardly be called free (Roberts 2005; Yergin 2011).

The electricity market, by contrast, is much freer. For most of the twentieth century, electricity was generated and distributed by big utility companies, many of them state owned. In most cases, customers, whether single households or large companies, held a

---

[4] See CNNMoney: http://money.cnn.com/magazines/fortune/global500/2012/full_list/.

direct contract with the utility. Beginning in the 1980s, many **electricity** markets throughout the world were liberalized and **deregulated**. Customers were no longer required to buy electricity from the **company** that generated it. This introduced more competition to the market, bringing down prices and giving consumers a vastly expanded choice. For example, many domestic power consumers in Europe can buy their electricity from small suppliers specialising in renewables. The electricity they use in their home is not the same physical entity that their supplier fed into the grid, yet a series of exchange mechanisms ensures that there is a direct correspondence between the two (Kopsakangas-Savolainen and Svento 2012).

5.3 The Quest for Energy Independence

Winston Churchill not only foresaw the rise of Hitler, he also anticipated the rise of oil. In 1911, he warned the House of Commons that Britain's future as a world power would depend on a reliable supply of oil. Just a few years previously, the British navy had begun switching from coal to kerosene fuelled ships. This allowed for faster and more powerful vessels, but introduced a new insecurity in terms of supply; instead of Welsh coal, British sea power now depended on Persian oil. Britain's global hegemony in the nineteenth century was built on coal, while the rise of the United States and Soviet Union to superpower status in the twentieth century was predicated on access to oil. The German and Japanese grabs for power in World War II were at least as much about energy resources as they were about territory. The Nazis understood that without secure oil resources Germany would not be able to replace Britain as Europe's dominant power. Their failure to capture the oil fields of the Caucasus signalled the end of the German dream of becoming the continental superpower.

By the 1970s, preoccupation with secure access to energy had migrated from the boardrooms and war rooms to the street. Hence U.S. President Richard Nixon's insistence on including the promise of "energy independence by 1980" in his 1973 State of the Union address. That was the year of the first oil crisis, and Nixon understood the political advantage of this promise. He deliberately modelled it on President John F. Kennedy's 1961 pledge to put an American on the moon. However, Nixon greatly underestimated the difficulty of ending his country's dependence on foreign oil. Whereas Kennedy's task depended largely on generous state funding and scientific ingenuity, Nixon's called for a transformation of the very fabric of American society. The mobility implicit in the American Dream was, and still is, tightly bound to hydrocarbons. The only way, then, to bring about 'energy independence' was either to develop viable alternatives or to discover vast new oil fields in the United States. Although the promise of energy independence has become a mainstay of U.S. politics into the twenty-first century, it is as far from realization today as it was in 1973 (Yergin 2011).

Every industrialised or industrialising country grapples with the problem of energy supply. Germany is the most recent country to make a promise reminiscent of Nixon's, in this case, to dispense with nuclear energy by 2022. While Germany's motives are more about safety than geopolitics, the challenge will be the same: to develop alternatives, discover new sources, and use energy more efficiently. In Germany's case, the departure from nuclear power is likely to lead – in the short-term at least – to a greater dependence on foreign energy resources, particularly Russian gas (Marquart 2011).

The promise of energy independence remains so elusive because, not unlike promises of full employment and uninterrupted growth, it is simply not realistic. Energy is the most globalised of all commodities, and all countries, importers and exporters alike, are bound to the markets. While the promise of energy independence continues to beguile voters, it has long since been abandoned by the captains of industry in favour of a more pliable concept: energy security. This entails an acceptance that energy systems are globally interdependent, and likely to become more so in the future, and implies a multifaceted approach to energy supply.

5.4 The Environmental Cost of Energy

As a child, I travelled with my family a few times a year from our home in the west of Ireland to visit my grandfather, who lived in a redbrick house in Dublin's inner city. Among the many lures of the city for a young bumpkin like myself were sweetshops, streetwise cousins, and the ubiquitous cacophony of traffic sounds. After a day playing outside in the lanes around my grandfather's house, we children returned home with blackened faces and hands. Dublin, in the early 1980s, was a city in which most households still burned coal in open fireplaces. I remember a light powdering of soot on every surface. Most of the grand old buildings in the city centre were black from centuries of air pollution, a reminder that the wealth of the industrialized West was built on a mountain of coal.

This story is mirrored in most cities in the industrialised world. The industrial expansion that began in the mid- to late nineteenth century had, by the mid-twentieth, thoroughly contaminated the air and water in most European and North American cities. The cleanup began gradually, spurred by dramatic events such as the Great London Smog of 1952, when a blanket of coal smoke hung over the city for a week, claiming 4,000 lives (McNeill 2010). Manufacturing companies enjoyed near impunity for decades, exemplified by the case of the Japanese Chisso Corporation, whose chemical factory pumped mercury into Minamata Bay in Japan for four decades, causing thousands of cases of brain damage and death (McNeill 2001).

Today, most cities in Europe, North America, and Japan are much cleaner than they were a generation ago. This change is partly because of environmental legislation and partly because the West has exported many of its dirtiest and most energy-intensive industries, including shipbuilding, textiles, paper and steel, to China, India, Mexico and other 'emerging' economies. Factories in those countries rely mostly on coal-fired electricity, generated without the filters and scrubbers that would be required by law in the West. Therefore, the relatively clean lakes, rivers, and air in the West have been achieved at the cost of poisoned rivers and smog-shrouded cities in the East.

Some argue that there is no way to avoid a dirty initial phase in economic growth, and that countries such as China, India, and Brazil are today going through an industrial revolution similar to that of Europe and North America in the nineteenth century. Energy consumption is so intertwined with economic growth that governments are wary of limiting the former for fear of stifling the latter. China continues to rely heavily on coal rather than gas or renewables because it is cheap and plentiful, and its economic success is built on quick and cheap production. Up to now, China has argued that economic growth trumps environmental protection, and that raising incomes takes priority over reducing smog. This is likely to change, as it did in countries that industrialised earlier, once the costs of pollution outweigh the economic gain.

5.5 Climate Change: The Great Leveller

For centuries the seafaring nations of Europe dreamed of an alternative sea route to the Far East that, heading west, would cut out the long voyage around the southern tip of Africa and across the Indian

Ocean. Between 1746 and 1846, ten separate expeditions tried and failed to discover a 'Northwest Passage' through the Arctic. The last of these became a cause célèbre in Britain when the two ships led by Sir John Franklin disappeared. Several further ships were lost searching for Franklin and his crew. In 1903–1906 an expedition led by Roald Amundsen finally made it through in a small shallow-bottomed ship, but the Northwest Passage remained impassable for commercial shipping for another century. In 2008, a Canadian shipping company, Nunavut Sealift and Supply, announced the start of year-round cargo services from Montreal to Western Canada. Some scientists now predict that, within the next forty years, John Franklin's frozen grave will be as busy a commercial shipping route as the Panama Canal (Stix 2006).

China, with characteristic pragmatism, is hedging its bets on energy. While rapidly expanding its coal power capacity, it has also invested heavily in hydropower. However, radical changes in weather have thrown a spanner in the works. In September 2011, many of China's rivers ran with less than 20 per cent of their normal water flow, and the nation's hydroelectric power output fell proportionally. The Chinese government has not provided an official opinion, but, according to Lin Boqiang, director of the China Center for Energy Economics Research at Xiamen University, "if climate change caused this year's water flow decreases, which I think it did, then its impact will be long-term. It will take a toll on China's hydroelectric output, and also push up the cost of using it" (Liu 2011).

There is no shortage of evidence, nor of anecdotes, pointing to a change in the Earth's climate. The term 'climate change' sounds oddly vague, 'change' being open to interpretation, and more suggestive of a political slogan than a scientific hypothesis. Yet ambiguity is at the very heart of the matter. The coalition of business interests, conservative politicians, and conspiracy theorists who deny or downplay climate change has latched on to this uncertainty, in much the same way that religious 'creationists' insist that evolution is only a 'theory'. The problem with predicting the exact nature of climate change is that so many variables are involved. While we have reliable data showing that the average temperature of the Earth's atmosphere and oceans is rising, no one really knows what effect this will have on climate, other than that it will change; the greater the rise in temperature, the more radical that change. But why is the Earth getting warmer, what is the cause, and what can we do to offset or 'mitigate' the change? In the end, it all comes down to our relationship with the sun.

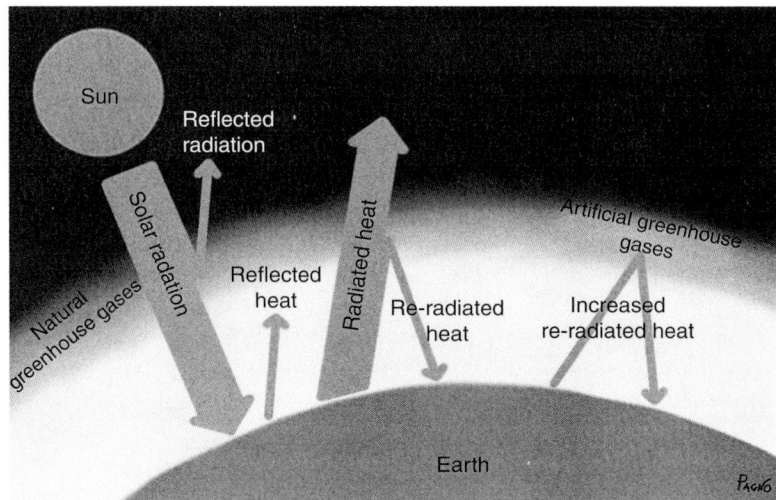

Figure 5.1. The greenhouse effect.

Solar radiation travels to Earth and breaks in through the atmosphere. Certain gases in the atmosphere immediately reflect a part of the radiation back into space. The rest is either absorbed by our atmosphere – its energy driving wind, waves, and weather – or reaches the ground and heats the Earth's surface. When it first enters the atmosphere, the solar radiation loses energy, its waves becoming longer (mostly in the infrared spectrum) and less intense. Having lost much of its energy, it is less able to penetrate the atmosphere on the way out than it was on the way in. This radiation is therefore trapped within the Earth's 'greenhouse' (see Figure 5.1).

As a student, I discovered firsthand the basic principle behind the greenhouse effect. At the end of a long Saturday night, my friend Garry and I were thirsty for more beer, but all the bars in town had closed. As we approached my flat, we passed a large warehouse yard belonging to a brewery. We paused before its unguarded gates and gazed inside, like hungry sheep into a lush meadow. I turned to Garry with an impish smile and he quickly guessed what I was thinking. He was a year older and his greater wisdom showed in his remark: "Steve, we could break in, but could we break back out again?"

Without the "greenhouse effect" our planet would have an average surface temperature of −18 degrees Celsius, instead of its actual +15 degrees. Water would exist, as on Mars, only in the form of ice. Indeed, the only place on Earth where life could exist would be surrounding thermal vents or geysers. Evolution of life, therefore, could not have progressed beyond very simple forms, such as bacteria, algae, and small crustaceans.

So, why is it that the greenhouse effect is usually spoken of as a problem? This comes from the fact that when most people speak about the greenhouse effect, what they really mean is the artificial greenhouse effect. The greenhouse gases (GHGs) – carbon dioxide (CO_2), methane (CH_4), nitrous oxide (N_2O) and ozone – are not pollutants in and of themselves; in fact, they are a natural part of the Earth's life cycle. Plants, for example, live by converting carbon dioxide into chemical energy – glucose – through photosynthesis. Indeed water, the elixir of life on Earth, is, in the form of vapour, the most potent of all GHGs.

Because of human activity, we have witnessed in the past 200 years a steady and accelerating increase in the quantity of GHGs in the Earth's atmosphere. This is because we have been burning fossil fuels to generate energy, cutting down forests for timber and to make way for cities and farms, and because those farms are more intensive and industrialized than ever before. These GHGs have become pollutants because the quantities emitted disturb the natural balance.

Scientists have been aware of atmospheric warming since at least the 1950s. However, it wasn't until the late 1960s that human interference with the natural climatic order became a subject of public debate. In 1968 the American biologist Paul Ehrlich wrote *The Population Bomb*, in which he warned of the dangers associated with increasing human population. Ehrlich's words were remarkably prescient: "the greenhouse effect is being enhanced now by the greatly increased level of carbon dioxide" (1968, 52). By the 1970s, theories such as this were no longer at the fringe; the scientific community was moving towards a consensus that human activity was causing the Earth to warm up. In 1979, the First World Climate Conference of the United Nations' World Meteorological Organization (WMO) spelled out the problem of global warming in much the same terms generally accepted today:

> We can say with some confidence that the burning of fossil fuels, deforestation, and changes of land use have increased the amount of carbon dioxide in the atmosphere by about 15 per cent during the last century and it is at present increasing by about 0.4 per cent per year. It is likely that an increase will continue in the future. Carbon dioxide plays a fundamental role in determining the temperature of the earth's atmosphere, and

it appears plausible that an increased amount of carbon dioxide in the atmosphere can contribute to a gradual warming of the lower atmosphere, especially at higher latitudes. Patterns of change would be likely to affect the distribution of temperature, rainfall, and other meteorological parameters, but the details of the changes are still poorly understood.[5]

By 1988, the international scientific community was devoting considerable energy to the issue of climate change. That year saw the foundation of the Intergovernmental Panel on Climate Change (IPCC), a joint venture by the WMO and the United Nations Environment Programme (UNEP). The IPCC's task has been to review the most recent scientific studies on climate change and to provide accurate and unbiased information to governments worldwide. The most recent IPCC findings, released in 2013, are unequivocal, confirming and specifying previous reports of changes to the global ecosystem. The atmosphere and ocean have warmed, the amounts of snow and ice have diminished, sea levels have risen, and concentrations of GHGs have increased. Evidence of human influence has grown and it is now considered extremely likely that human influence has been the dominant cause of the observed warming since the mid-twentieth century.

Atmospheric concentrations of carbon dioxide, methane, and nitrous oxide have increased to levels unprecedented for at least the last 800,000 years. Carbon dioxide concentrations have increased by 40 per cent since pre-industrial times (from 278 parts per million in 1750 to 390 parts per million in 2011[6]), primarily as a result of fossil fuel emissions and changes in land use.

Provided the growth of GHG emissions remains constant, the average surface temperature of the Earth will increase and new weather records will be recorded at ever-shorter frequencies, as has been the case for the last thirty years. A range of knock-on effects is expected. The ice caps will continue to melt, resulting in a rise in world sea levels of more than 50 centimetres before the end of the twenty-first century. This could submerge many coastal cities, such as Venice, New York, Athens, and Tokyo. The oceans will become more acidic, which

---

[5] Declaration of the World Climate Conference of the World Meteorological Organization (https://www.wmo.int/pages/prog/wcp/documents/WMO-540_PartB_Annexes.pdf).

[6] "Parts per million" is a term commonly used to quantify levels of air pollution. It literally means the number of molecules of a given pollutant in one million molecules of air.

will have major impacts on marine life, and on land many plants and animals will become extinct. Because of the increased temperatures, deserts will expand, and freshwater – for drinking and irrigation – will become scarcer. As a result, there will be less land available for agriculture, and therefore less food produced (IPCC 2007, 2013).

Climate change sceptics often insist that there is nothing unnatural about global warming, since it happened before. This is true. Lee Kump of Pennsylvania State University recently confirmed that the current period of global warming is not the first in the Earth's 4.5-billion-year history. Thermal peaks occurred 100 and 55 million years ago, most likely a result of GHG emissions from volcanic eruptions. The most important difference between these and the current phase of global warming is the speed. The first occurred over tens of millennia, giving life-forms time to evolve and adapt to the new conditions, whereas modern warming is happening over a period of just a few centuries, far too fast for evolution to keep pace (Kump 2011).

While the nations of the world can largely agree that climate change is a major threat to humanity, there is considerable disagreement on the best way to deal with it. While the wealthier industrialised countries are thinking (and so far doing little more than that) about lowering GHG emissions, the energy-hungry economies of the 'developing world' are inclined to regard emissions reduction as a luxury they cannot yet afford. Viewed from Europe and North America, China's embrace of coal may seem irresponsible, but its economists can convincingly argue that its primary responsibility is to raise the living standards of its people, and this is only possible with a huge increase in power production. China recently overtook the United States as the world's largest emitter of GHGs, yet its per capita emissions are still far lower. China also only began industrialising in the late twentieth century, so its historic carbon footprint is still far smaller than that of Western economies. Zhang Guobao, director of China's National Energy Administration, argues that the pressure the international community places on his country is unfair, since China, like other developing nations, is only following in the footsteps of the rich West (Biello 2010).

Climate change presents the community of nations, itself a recent concept, with its greatest challenge yet. Global warming, which recognises no borders and does not distinguish between rich and poor nations, demands a concerted act of creative cooperation between all peoples of the world.

5.6 The Green Movement

Environmentalism, as we know it today, traces its origins back to the late nineteenth century when a movement began in the industrial societies of Europe and North America to protect natural landscapes from the ravages of industrialisation. However, it was not until the 1970s that a distinct and politically discrete environmental movement emerged. Spurred by several conspicuous incidents of environmental pollution and influential books, such as Rachel Carson's *Silent Spring*, which exposed the dangers of pesticide use, environmentalism became a major public concern (Carson 2002). Earth Day in 1970, organised by a group of Harvard students, was the first major event to bring together many of the issues that form the modern Green manifesto: air and water pollution, industrial waste, overpopulation. Following the example of the anti–Vietnam War movement, 'teach-ins' were held in different cities across the United States. As many as 20 million Americans took part. In the aftermath of the event, U.S. Congress passed its first Clean Air Act and established the Environmental Protection Agency. Throughout the 1970s political parties with a primarily conservationist or environmental emphasis emerged throughout the industrialised world (Yergin 2011).

The catalyst for a global conversation about the future of human development came from an unlikely source: a report commissioned by one of the world's first 'think tanks', the Club of Rome. The authors, environmental scientists at the Massachusetts Institute of Technology, used a computer model to predict the collapse of human population and the global economic system if human behaviour did not radically change. *The Limits to Growth* struck a chord in Western societies, with 12 million copies sold since its publication in 1972, an unprecedented achievement for a scholarly publication (Meadows et al. 1974).[7] Its success reflected a growing conviction in the industrialized world that the economic success and 'development' which had transformed standards of living in Europe and North America since the Second World War were not sustainable.

---

[7] The report focused on five areas: population, food production, industry, pollution, and resource consumption. The report's dire predictions about population growth have not been realised. Global population has roughly doubled since the *Limits of Growth* was published, and while poverty and malnourishment remain major problems, proportionally no more so than in 1972.

5.7 Sustainable Development

At the turn of the twentieth century, many in the industrialised countries were expressing concern about the social and environmental costs of economic growth. However, two world wars and an economic depression drew attention to more immediate matters. The concept of 'sustainable extraction' first emerged from the conservation movement and was applied to the regulation of hunting and fishing. In 1953, the American conservationist Henry Fairfield Osborn wrote that "man is becoming aware of the limits of his earth" (Adams 2004, 154). Space exploration, and particularly the first images of the Earth brought back by the Apollo missions, may also have triggered a change in attitude to the global environment. Frank Boormann, commander of the Apollo 8 mission wrote, "When you are privileged to view the earth from afar … you realise that we are really, all of us around the world, crew members on the space station Earth" (Adams 2004, 154). The image of the fragile blue planet suspended in dark space became an icon of the environmental movement.

Responding to the growing concerns about the future of 'development', the UN established a World Commission on the Environment and Development in 1984. Known as the Brundtland Commission after its chairperson, former Norwegian prime minister Gro Harlem Brundtland, the commission spent three years gathering data for its report, entitled *Our Common Future* (United Nations 1987). While treating environmental degradation with the same sense of urgency as war and poverty – as global threats to human existence – the report softened the Cassandra warnings of *The Limits to Growth*. Rather than decry economic growth per se, the report stressed the need to change the quality of growth by incorporating environmental concerns in economic decision making. In doing so, it proposed an alternative paradigm: sustainable development. It defined this as "development that meets the needs of the present without compromising the ability of future generations to meet their own needs" (United Nations 1987, 41). This marked the moment when environmentalism entered the boardrooms of business and government.

Building on *Our Common Future*, the UN convened a conference on development and the environment in Rio de Janeiro in June 1992. No such event had taken place before. More than 20,000 people, representing governments, industry, and a variety of nongovernmental organisations, participated directly or indirectly in what was somewhat grandiosely dubbed the 'Earth Summit'. One hundred and forty-four heads of state or government attended, as did 10,000 journalists. Perhaps its sheer scale raised expectations too high, for the Rio Summit did not produce the bold Earth Charter that was hoped for. However, two specific and legally binding agreements did emerge from the summit: the Convention on Biological Diversity and the United Nations Framework Convention on Climate Change (UNFCCC), an international treaty to cooperatively consider what the international community can do to limit climate change, and to cope with whatever impacts were, by then, inevitable.[8] In 1997, negotiators followed up with the Kyoto Protocol, which, operative since 2005, committed industrialized countries to reducing their collective emissions of GHGs by 5.2 per cent by 2012, compared with 1990 levels.

Richard Benedick, who negotiated the Montreal Protocol (the international treaty that phased out the use of certain ozone-depleting substances) on behalf of the United States, warned that progress in limiting GHG emissions would be slow. Whereas the Montreal Protocol was dramatically successful in preventing further ozone depletion, one of the biggest environmental concerns of the 1980s, this was a far easier problem to solve than man-made global warming. The global ban on the most aggressive ozone-depleting substances (mostly chlorofluorocarbons or CFCs, gases used in aerosols and refrigerators) affected only a few industries, which quickly came up with less harmful replacements. By contrast, almost every human activity currently involves the consumption of energy, and the vast majority of our energy comes from carbon.

As Benedick predicted, not only have we failed to reduce carbon dioxide emissions since Rio; we have increased the pace of emission. Humans pumped 22.7 billion tonnes of carbon dioxide into the atmosphere in 1990.[9] By 2010, that had increased to 33 billion tonnes. This is not as bad news as it may at first seem. The world economy has been transformed since 1990, as new economic giants, such as China, India and Brazil, have emerged. To reflect the differences between the

[8] The Convention on Biological Diversity, signed by 150 government leaders at the 1992 Rio Earth Summit and entered into force on December 29, 1993, is dedicated to promoting sustainable development and recognizes that biological diversity is about more than plants, animals, and microorganisms and their ecosystems – it is also about people and food security, medicines, fresh air and water, shelter, and a clean and healthy environment in which to live. It has three main objectives: (1) the conservation of biological diversity; (2) the sustainable use of the components of biological diversity; (3) the fair and equitable sharing of the benefits arising out of the utilization of genetic resources.
[9] The baseline year under the UN Framework Convention on Climate Change.

old and the new economic powers, the Kyoto treaty introduced the concept of 'common but differentiated responsibilities'. This means that wealthier countries, which have historically been responsible for the largest share of carbon dioxide emissions, should bear a greater responsibility for reducing their emissions. These countries as a group are on track to exceed the Kyoto goal with an overall reduction of 7 per cent, but this is largely attributable to the demise of the Soviet Union and the more recent financial crisis (Tollefson et al. 2012).

The Rio Summit took place just two years after the end of the Cold War. It was a time of great optimism, tinged perhaps with a certain naïveté, since no precedent yet existed for truly global action towards a common long-term goal. On balance, the climate treaties and structures that emerged from the Rio Summit, though not a dramatic success, have achieved some important results. They encouraged investment in climate science, raised global awareness of climate change, and spurred governments to take the first steps towards dealing with specific problems, such as desertification, sustainable agriculture and tropical deforestation, and to consider innovative solutions, such as carbon markets and transfers – of money and technology – to poor countries.

Two decades after Rio, the problem is more acute and the solution as elusive as ever. What progress there is, is slow and difficult to measure. In the words of David Victor, director of the Laboratory on International Law and Regulation at the University of California, San Diego, "Plausibly we are a little better off than if we didn't have all of this diplomacy, but the evidence is hard to find" (Tollefson et al. 2012).

5.8 The Nuclear Option

Nuclear energy is clearly not renewable, since the reserves of nuclear fuel are finite. Nor would many see it as 'green' because of the risks associated with waste and accidents. However, some notable voices, including Jesse Ausubel, co-organizer of the first UN World Climate Conference, environmentalist James Lovelock, author of *The Revenge of Gaia* (2007), and former Greenpeace Director Stephen Tindale, support the view that nuclear power needs to be embraced as part of a sustainable approach to energy production, at least in the short to medium term. Ausubel goes as far as to dismiss renewables such as wind, water, and biomass as 'boutique fuels' that are not truly green since they would "cause serious environmental harm" if produced at a scale that would contribute importantly to meeting global energy demand (Ausubel 2007).

Civilian nuclear technology was born out of military research by the Allied powers after World War II, and to this day the nuclear power sector has difficulty shedding its association with weapons of mass destruction. The USSR, Britain, and the United States commissioned the first nuclear reactors for civilian electricity generation in quick succession between 1954 and 1957. Currently there are roughly 430 nuclear power plants in operation in thirty countries, producing 13 per cent of the world's electricity. However, only ten countries rely on nuclear power to meet more than one-third of their electricity needs, and only in three countries (France, Belgium, and Slovakia) is the majority of electricity nuclear-generated. Nine countries are known to possess nuclear weapons (IAEA 2010, 2012).

The relative decoupling of nuclear power from weapons is largely thanks to the work of the International Atomic Energy Agency, which was set up in 1957 to promote the peaceful use of nuclear energy, and the International Non-Proliferation Treaty, ratified by all but four of the world's nations.[10] Yet despite these safeguards and controls, there remains a danger that civilian nuclear power could serve as a cloak for the development of nuclear weapons. Although the uranium fuel required for a weapon is very different from that used in a reactor, the process for creating both is the same (IAEA 2012; Yergin 2011).

There are good reasons both to embrace and reject nuclear power. The big positive is that, according to the IEA, nuclear power has the "capability to deliver significant amounts of very low-carbon base-load electricity at costs stable over time" (IEA 2010a, 21). In other words, a nuclear plant is cheap and clean to run, in the conventional sense. Nuclear power generation is a mature technology, and the fuel costs are lower than for other conventional power sources.[11] This means that nuclear power is far more resistant to market fluctuations than coal or gas.

However, nuclear power plants cost a lot of money and take a long time to build, and, therefore, they take a long time to recoup

---

[10] India, Pakistan, North Korea, and Israel.
[11] Total nuclear fuel costs, which include mining and milling of raw uranium, as well as enrichment and fabrication of nuclear fuel, come to only 0.7 U.S. cents per kilowatt-hour, or between 10 per cent and 20 per cent of the overall costs of nuclear electricity, compared with 25–35 per cent for coal and 60–70 per cent for gas (IEA 2010a; WNA 2005).

their initial costs.¹² Moreover, the costs of nuclear waste disposal and plant decommissioning are difficult to calculate. When Britain privatized its electricity industry in the 1980s, there were no takers for the nuclear power plants (McNeill 2000, 312). They remained in state hands until 2009, when they were bought by the French state utility, Électricité de France. Uncertainty, both in terms of costs and safety, makes nuclear power a very unattractive bet for private investors, who generally want to see a return on investment within ten years. As a result, governments often step in with subsidies, financing, or loan guarantees.

If the Deepwater Horizon oil disaster in the Gulf of Mexico was almost enough to bankrupt BP,¹³ imagine the costs of cleaning up after a major nuclear accident. Major nuclear accidents are rare (Sellafield, Three Mile Island, Chernobyl, and Fukushima are the only ones to date),¹⁴ but this risk is reflected in high insurance costs. The overall cleanup and compensation costs of the Fukushima Daiichi catastrophe have been estimated at more than US$100 billion. As a result, the company was nationalised and the Japanese state took over the costs of the accident (Greenpeace 2012, 2013). There is also the problem of nuclear waste, which remains highly radioactive for tens of thousands of years. The challenge of finding a place to store that waste safely for such an immense period of time has so far proven overwhelming.

There was steady growth in the nuclear power sector from the mid-1970s to the mid-1980s, roughly from the time of the first oil crisis in 1973 to the Chernobyl disaster in 1986. After Chernobyl, the nuclear industry went into decline. Many reactors that had been ordered or partly built were never completed, and some countries (notably Italy and Sweden) opted to decommission their existing nuclear plants (Cooper 2009). Nuclear power began to make a comeback in the first decade of this century, with thirty-nine new plants connected to the grid between 2000 and 2010, most of them in Asia. Even in Europe, the memory of Chernobyl appeared to have receded sufficiently for nuclear energy to be reluctantly embraced as the quickest route to

---

[12] According to a 2003 U.S. congressional budget office report, capital costs per kilowatt of capacity for nuclear power plants are 40 per cent to 250 per cent above those for electricity plants using gas and coal. Construction time for nuclear power plants varies from six to twelve years (Cooper 2009).

[13] As of July 2012, the oil spill had cost BP close to US$40 billion, more than one and a half times the company's annual profits (*Guardian*, July 31, 2012).

[14] These are the only accidents that have been categorized at Level 5 or higher on the IAEA International Nuclear Events Scale.

reach the carbon-reduction targets agreed on at Kyoto. Had it not been for an earthquake off the coast of Japan, which triggered a tsunami and led to the partial meltdown of the Fukushima Daiichi plant in March 2011, we would now probably be experiencing a renaissance of nuclear energy.

6
The Price of Energy Consumption

6.1 Energy and the Environment

> It was a town of machinery and tall chimneys, out of which interminable serpents of smoke trailed themselves for ever and ever, and never got uncoiled. It had a black canal in it, and a river that ran purple with ill-smelling dye, and vast piles of building full of windows where there was a rattling and a trembling all day long, and where the piston of the steam-engine worked monotonously up and down, like the head of an elephant in a state of melancholy madness. (Dickens 1995)

Charles Dickens called the setting of *Hard Times* Coketown. Like many English towns and cities in the nineteenth century, it was defined by the dirty process of producing coke, a coal by-product used to make steel. England was the cradle of the Industrial Revolution, and by Dickens's time it was clear that this was a very messy baby. To appreciate the environmental impact of the world's first wave of industrialisation, we need to look outside Europe and North America, to countries like China, India, Brazil, and Indonesia that are riding the latest wave. Manchester, Duisburg, Pittsburg and Turin are no longer foul-smelling and coal-blackened, to a large extent because Beijing, New Delhi, São Paulo and Lagos are. As in Dickens's time, the environmental footprint of human societies is intimately linked to our consumption of energy.

Long before our energy-devouring societies began to pollute the planet's air and water, agriculture put the stamp of humans firmly on the land. Intensive agriculture initially supported a growing rural population, but from the eighteenth century, farm mechanization pushed, and urban industry pulled, millions of people into cities. The mechanization of agriculture changed humankind's relationship with the land. We went from a sense of dependence on its bounty to one of almost infinite control over its 'products'.[1] The progression from rural, decentralized, and low-energy societies to urban, centralized and high-energy ones has largely run its course in wealthy regions, but is still ongoing in most of the world.[2]

Urbanisation is both a cause and an effect of changes in how land and energy are used. For millennia, humans relied on warm summers, mild harvest seasons, animal manure, and crop rotations (to let the soil recover nutrients) to ensure a plentiful supply of food. The injection of additional energy into agricultural systems began with animal power and manure, later involving fossil fuels with the use of tractors, mechanized irrigation (canals and water pumps), and artificial fertilizers. Between 1940 and 1990, the global use of chemical fertilizers grew from 4 million to 150 million tonnes per year. Like so many other innovations, fertilizers have been a mixed blessing. They have greatly boosted crop yields, providing food for an estimated 2 billion people, but have also encouraged monoculture, reduced biodiversity, contaminated watercourses and significantly impaired the quality of the soil.

The history of human civilisation is recorded in land. Satellite images have only been available for a few decades. If we were able to compare images of the Earth's surface taken from space over a much longer period, we could trace the advance of technological civilisation. Over the course of five millennia, humans have carved out fields, felled and burned forests, dammed and directed rivers, drained marshes, and dug mines. Even a millennium ago, the impact of human activity would have been visible from space, thanks to forest clearances in Europe and Asia. Two hundred years ago vast patchworks of enclosed fields, open-cast mines several kilometres in diameter, and the first urban sprawl would have been visible. Since then, humankind's

---

[1] Hunting and gathering societies could sustain fewer than one person per square kilometre. Pastoralism raised the sustainable human settlement density to two individuals per square kilometre, and farming up to ten people per square kilometre. In 1800, only 3 per cent of the world's population lived in towns and cities. By 1900 the share had risen to 15 per cent, and in 2008, for the first time, the majority of the world's human population was urban (Smil 2008).

[2] In most countries of North America and Europe the urban population is above 75 per cent. In China, a slim majority (55 per cent) still live in villages, a major transformation since 1950, when 89 per cent of Chinese lived on the land (Kojima 1995).

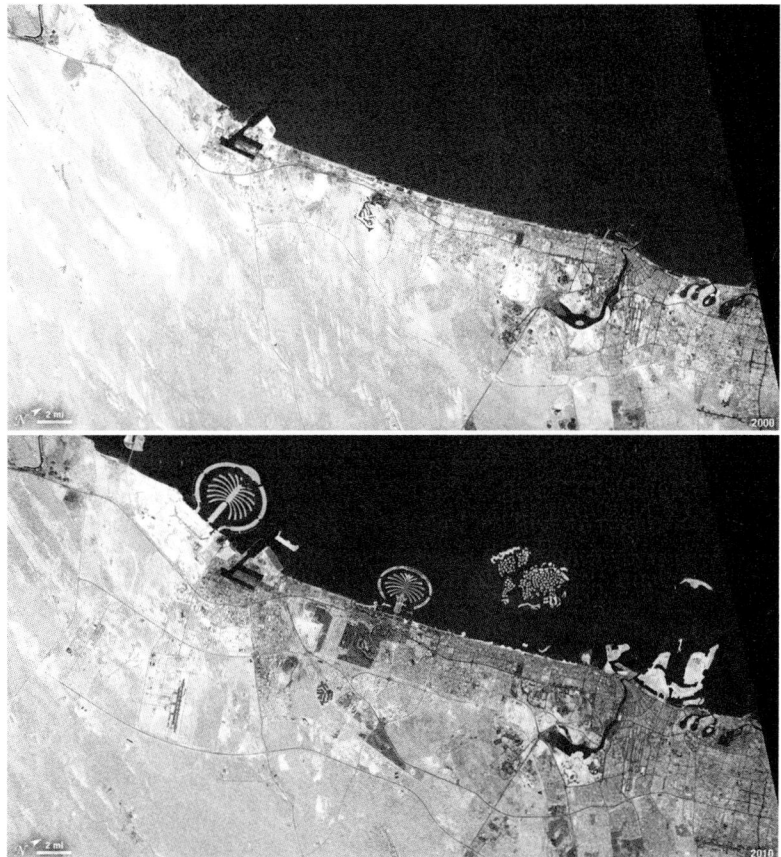

Figure 6.1. In 2000, Dubai was an isolated triangle of urban development in the Arabian Desert. In just ten years the view from space was transformed as Dubai boomed on a mixture of tourism, trade, and oil production. Today, city blocks are surrounded by green land irrigated with desalinated seawater and the coastline is dotted with artificial islands.
Source: NASA/USGS.

topographical footprint has deepened and broadened. Today more than 30 per cent of the Earth's land surface is appropriated for human needs (Andrews et al. 2011).

Land Impacts

The impacts of energy use on land may be grouped into three categories. The first involves the least impact and includes coal, natural

Figure 6.2. A strip coal mining operation in Australia.
Source: Stephen Codrington at Wikimedia Commons.

gas, and nuclear power, through the extraction of highly concentrated resources and their conversion to energy or more refined fuels in power plants or refineries. Of the alternative energy resources, only geothermal and solar thermal belong in the same low-impact category; geothermal because it uses gas- and oil-drilling technology to extract hot water, and low-temperature solar thermal systems because they are usually located on rooftops and therefore do not appropriate additional land. The footprints of high-temperature solar thermal systems (CSP) are also small, with land impacts similar to coal and natural gas.

Petroleum, hydropower, solar photovoltaics (PV), wind power and low-temperature solar thermal systems for district heating belong in the second category, involving medium-level land impacts. The vast network of pipelines and refineries, the extraction of tar sands in Alberta, the reservoir of the Three Gorges Dam, and the vast arrays of solar PV panels needed to generate electricity on an industrial scale are among the manifestations of these impacts. Wind power also has a considerable land impact because turbines are usually placed a few hundred metres apart to avoid mutual interference. However, this is mitigated by the fact that wind farms can coexist with either cropland or pastoral farming.³

The third, highest-impact category is reserved for biofuels. Because of the low efficiency of photosynthesis (Section 1.2), a large area of land is required to produce biomass fuels, whether in the form of timber for direct combustion or an energy crop for alcoholic fermentation.⁴ Sugarcane ethanol is the most efficient of the biofuel crops, yet to meet global primary production we would need the entire land area currently used by humans, and using biofuels derived from soy we would need the entire land surface of the Earth (see Figure 6.3). Biofuels will eventually occupy an important niche in the energy mix, but physical limits will prevent them from becoming a globally dominant energy solution (Andrews et al. 2011).

Air Impacts

When wood, coal, or oil is burned, the most evident manifestation is the smoke. It comprises tiny particles that can penetrate the lungs, enter the bloodstream, causing heart disease, lung cancer, asthma and respiratory infections.⁵ More than 2 million people die every year from inhaling particles floating in indoor and outdoor air (WHO 2011). In both developed and developing countries, the biggest contributors to air pollution are exhaust fumes from transport and industries, combustion of wood for cooking and heating, and conventional fossil fuel-based power plants. In developing countries, traditional burning of

---

[3] Since the energy is generated above the ground, the only impact at ground level is the area taken up by the base of the tower.

[4] Current estimates of the ranges of land use intensity (in square kilometres per TWh-year) associated with various energy sources, based on the expected state of technology in the year 2030, are nuclear power (1–3), geothermal (1–14), coal (3–17), solar thermal (10–20), natural gas (1–36), solar PV (21–53), petroleum (1–88), hydropower (16–92), wind (65–79), sugarcane ethanol (220–350), corn ethanol (320–375), cellulosic ethanol (120–750), electricity from biomass (433–654), and soy biodiesel (780–1,000). These estimates take into consideration the land needed for resource extraction, processing, conversion, and waste storage (Andrews et al. 2011; McDonald et al. 2009).

[5] Particulate matter, or PM, is a complex mixture of extremely small particles and liquid droplets. PM is usually grouped into two categories: (1) inhalable coarse particles found near roadways and dusty industries, which are smaller than 10 and larger than 2.5 micrometers in diameter, and (2) fine particles, found in smoke and haze, less than 2.5 micrometres in diameter.

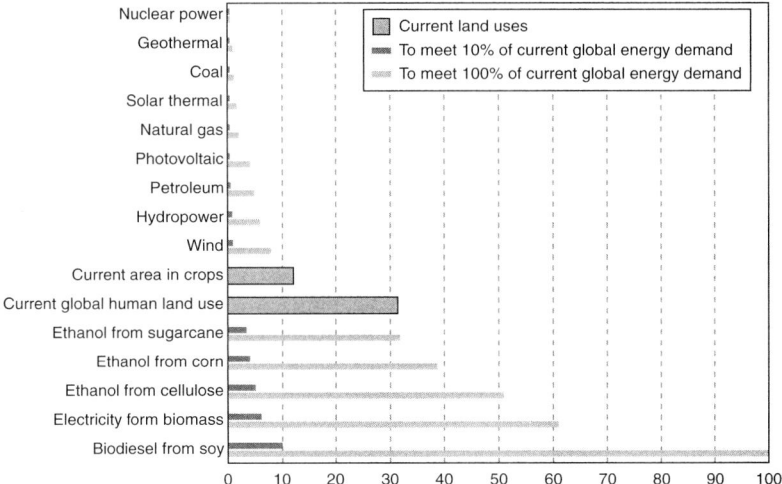

Figure 6.3. Land requirements by energy resource type to meet 10 percent and 100 percent of 2010 global energy demand, based on current conversion technologies. Data are compared with current human land use. This shows that biofuels are by far the most land-intensive of all energy sources.
Source: Andrews et al. (2011) (modified).

solid fuels (e.g., cooking or space heating with wood or coal) exposes many people to indoor smoke fumes. In addition to particles, carbon dioxide, sulphur dioxide, nitrogen oxides, metals, and various kinds of dust are released by the combustion of wood, coal, or oil. Each has different impacts on human health, ecosystems and climate. The burning of natural gas is less hazardous since it releases nitrogen oxides and carbon dioxide but in lower quantities than the burning of coal or oil.

Most exposure to air pollutants is directly or indirectly linked to energy production. Household environmental exposures, including indoor air pollution from the burning of solid fuels for cooking and heating, is highest in less industrialized countries.

In China, the world's fastest-growing major economy, the problem of air pollution is reaching acute levels. Under the socialist Huai River policy, from 1950 to 1980, the Chinese government provided free coal for heating homes and offices in the colder regions north of the Huai River. The situation was different south of the river, where budget constraints prevented the provision of free coal. A recent study examined pollution and mortality data from sites across the country for the years 1981 to 2000 (Chen et al. 2013). Concentrations of

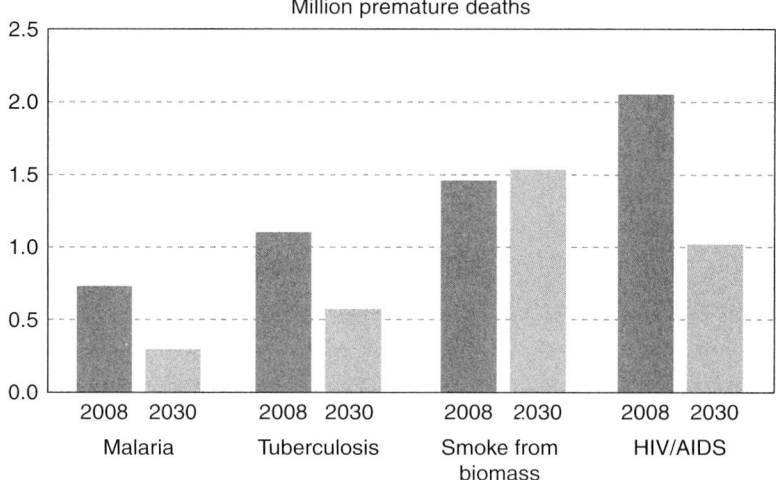

Figure 6.4. Premature deaths from household air pollution and other diseases in 2008 and projected for 2030.
Source: Sathaye et al. (2011) (modified).

suspended particulates north of the river were found to be 55 per cent higher than south. The inhabitants of the northern region paid for their cheaper energy with their lives, as it was shown that an increased incidence of cardiorespiratory mortality led to an average reduction in life expectancy of 5.5 years.

Exposure to domestic fumes is recognized as one of the biggest causes of early death in developing countries, and it is projected to exceed other major causes of premature deaths, such as AIDS, tuberculosis and malaria, by 2030. Women and children, who spend more time indoors, and the poorest segment of the population are most affected (Fischedick et al. 2011; Sathaye et al. 2011; WHO 2011).

Water Impacts

Humans consume 4,000 cubic kilometres of water every year, more than the entire volume of Lake Victoria. Most of this (2,800 cubic kilometres per year) is claimed by agriculture (Smil 2008). The energy industry accounts for roughly one-fifth of global water consumption (IEA 2012a). Some of this is used in extraction (water pumped underground to push out oil, coal or gas), some in refining, but most in power plants as a coolant and as the vector for steam turbines (see Figure 6.5). Hydroelectric

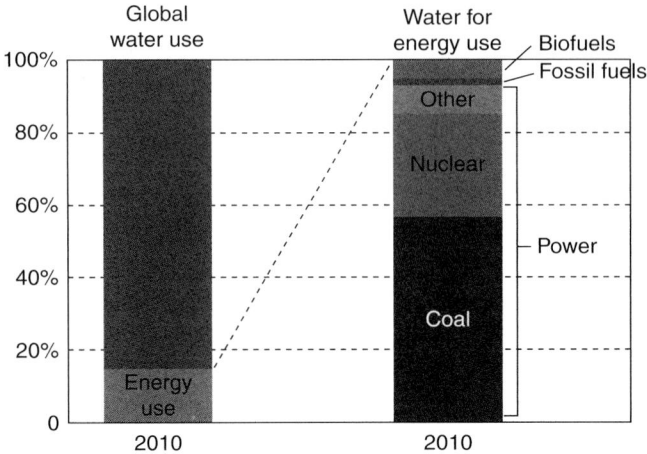

Figure 6.5. Water supply is as integral to the electricity sector as the fuels that are burned. The power sector's water needs are likely to grow, making water an increasingly important factor affecting the viability of energy projects.
Source: IEA (2012a) (modified).

schemes are also prodigious water users. More than half of the water diverted by hydroelectric dams is wasted (either through evaporation or dissipation into groundwater) before reaching hydro turbines.

Human control and use of water resources on a grand scale invariably entails pollution. The energy industry is the source of various forms of water contamination, ranging from acid mine drainage (the flow of acidic water from coal or metal mines into rivers and groundwater) to oil spills. Most of these releases have only localized impacts, and small oil spills are usually quickly mitigated by evaporation and the action of microbes. Large oil spills following large tanker accidents or offshore well leaks damage local ecosystems by killing fish and birds and, less obviously but more profoundly, by coating the ocean floor with slick that prevents oxygen from permeating the sediments that support the bottom of the marine food chain.

Finally, we must consider the energy cost of a water supply, which varies from 1 to 4 kilowatt-hours per cubic metre, depending on local conditions. When desalination is required, for example, in countries such as the Maldives or Saudi Arabia, where demand for freshwater far outstrips the natural availability, the energetic cost may be as high as 24 kilowatt-hours per cubic metre (Smil 2008).

Table 6.1. *Water demands of different activities in the energy industry compared with the global demand of agricultural crops*

Sector	l/GJ
Global crop harvest	150,000 l/GJ
Coal mining at surface	2–5 l/GJ
Coal deep mining	20 l/GJ
Oil extraction	5–10 l/GJ
Oil recovery from Alberta tar sands	30 l/GJ
Oil recovery with steam injection	500–600 l/GJ
Natural gas extraction	300 l/GJ
Coal cleaning	20–50 l/GJ
Oil refining	200 l/GJ

Note: Figures are in litres used per GJ of energy produced.
Source: Smil (2008) (modified).

6.2 The Great Polluters: Coal, Oil and Gas

Coal mining affects the environment by disturbing layers of soil and rock and exposing them to rain or groundwater. This may lead to acid mine drainage, when water mixes with the exposed rock, absorbing toxic levels of minerals and heavy metals. This water may then leak out of the mines to contaminate groundwater and streams, affecting soil, plants, animals, and humans.

For the first fifty years of the 'oil era', drilling for oil was an extremely dirty business. The environmental destruction, both on land and in water, of early oil discoveries in Azerbaijan (then part of the Russian Empire), Texas, Oklahoma, Mexico and Venezuela was enormous. Spills, leaks, blowouts and fires were part and parcel of oil extraction well into the twentieth century. The oil business transformed the coastal rainforest of Northern Veracruz, Mexico, into a poisonous oily morass in just twenty years. Something similar happened at Lake Maracaibo in Venezuela.

By the middle of the twentieth century, the oil industry had cleaned up its act considerably, as a result of the pressure of environmental regulations such as the 1924 Oil Pollution Act in the United States, and because of improved technology such as blowout preventers. However, oil extraction continues to take a heavy toll on the Earth's land and water resources, particularly in countries where environmental regulations are lax. A case in point is the Niger delta in

West Africa. Ever since oil was discovered there in the 1960s, serious pollution has followed apace. By the 1980s, much of the land and fisheries of the delta were heavily polluted. In 1992, the United Nations declared it to be the world's most endangered river delta (McNeill 2000, 304).

The site of extraction is not the only area impacted by oil production. Oil is mostly transported via two routes: by sea in tankers or over land through pipelines. As trade in oil has grown, so too have the ships used to transport it. Up to the end of the Second World War, oil tankers typically carried between 10,000 and 20,000 tonnes of crude oil. By the 1970s, the new class of 'ultra large crude carriers' had capacities exceeding 500,000 tonnes (Smil 1994). Even though improved technology meant that ships were more seaworthy than ever, the sheer size of modern tankers, plus the number of tankers plying the oceans at any given time, makes them a serious threat to the coastal and marine environment. The largest ship-based oil spill to date was the 276,000 tonnes of crude from the Atlantic Empress near Trinidad in 1979. Though this released nearly eight times as much oil as the Exxon Valdez spill off Alaska a decade later, it failed to catch headlines in the same way because most of the oil did not reach land.

Although more oil is traded today than ever before, in the last two decades spills from oil tankers have become far less frequent and, where they occur, less serious, thanks to improvements in ship design and safety regulations, and to the ever-expanding networks of oil pipelines. Pipelines might seem a more environmentally sound way of transporting oil, yet they are also more susceptible to spillages because of sabotage or theft. Secondly, migrating animals, from hedgehogs to caribou, experience oil and gas pipelines as barriers, preventing them from moving to wintering or breeding areas. For that reason, some modern pipelines are elevated to allow animals to pass underneath (see Figure 6.6).

The melting of arctic ice, largely the result of man-made global warming, has opened up the prospect of drilling for oil and gas in the polar regions of Alaska, Canada and Russia. The challenges of ensuring safety in such harsh conditions are numerous: drilling is limited to a few months per year; pack ice and icebergs complicate or prevent shipping; oil behaves differently at low temperatures; and the logistics of any cleanup operation become far more complex. A report commissioned by the U.S. government following the Deepwater Horizon accident warned that "many of the challenges emerging in Arctic oil and

Figure 6.6. The Trans-Alaska oil pipeline with a moose in the foreground. Because pipelines may block the migration routes of animals, many are now elevated.
Source: AK Smith at Wikimedia Commons.

gas development decision making are beyond the ability of science alone to resolve" (Holland-Bartels and Pierce 2011, 4).

6.3 Unconventional Sources of Conventional Pollution

In recent years, estimates of the world's reserves of oil and gas have been revised upward, mainly because of the inclusion of unconventional or 'tight' resources such as tar sands and shale gas (Chapter 2). The pollution from the extraction and processing of tar sands is a matter of growing concern (Timoney and Lee 2009). This is by far the most invasive form of oil extraction. The boreal forest and underlying peat are first scraped away to expose the tar sand deposits. Then, depending on the method, either steam, hot water, pressurized air or solvents are injected into the sands, allowing the petroleum to flow off.[6] This

---

[6] About two tons of tar sands are required to produce a single barrel of oil (roughly one-eighth of a tonne) from shallow deposits (less than 80 metres deep). For deeper deposits, superhot pressurized steam is pumped underground to melt out the bitumen, a process that requires less water but far more energy, resulting in greenhouse gas emissions 2.5 times higher than those from mining shallow deposits (Biello 2012b).

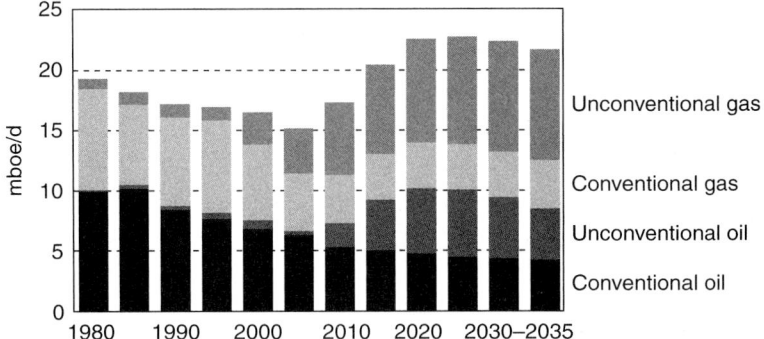

Figure 6.7. U.S. oil and gas production over the last thirty years and projection to 2035 (in million barrels of oil equivalent per day). This surge in unconventional resources will have implications well beyond the United States.
Source: IEA (2012a) (modified).

takes place either in an extraction plant (to which the sands are transported using giant strip-mining machines) or in the ground (known as in situ extraction).

The water and sludge that are left over from tar sand extraction, containing high levels of heavy metals and chemicals known as polycyclic aromatic hydrocarbons (PAHs), are stored in tailing ponds, which may leak. The levels of PAHs recently found in Alberta's lakes are two to twenty times higher than before tar sand mining began. PAHs and heavy metals are known to cause cancer and other health problems in animals and humans. According to Peter Hodson, a biologist at Queen's University in Ontario, "hydrocarbon pollution in the Athabasca River watershed is entirely caused by natural erosion of bitumen and is no longer tenable" (Biello 2013c).

Additionally, oil from tar sands is amongst the most greenhouse-gas-intensive forms of petroleum. The U.S. Environmental Protection Agency estimates that life cycle emissions from tar sands are about 80 percent greater than the emissions from crude oil. According to the renowned climatologist James Hansen, using resources such as tar sands will mean "game-over for climate change" (Biello 2012b, 2013e, 2013d, 2013a; Hodson 2012; Kurek et al. 2012; Sierra Club 2012).

Shale gas is equally, if not more, contentious than tar sand oil extraction. Up to now the most commercially viable method of extracting this natural gas trapped in shale rock has been to inject water under high pressure, combined with sand and a cocktail of chemical

Figure 6.8. A tar sand mine in Alberta (Canada) in 2006. The extraction site is to the left of the image, the plant to the right, and the sulphur stockpiles in the middle.
Source: TastyCakes/Jamitzky at Wikimedia Commons.

agents, thereby blasting pathways into the shale, allowing the gas to escape. This method, known as hydraulic fracturing or 'fracking' has only been widely applied in the last ten years, but it has already transformed the U.S. natural gas market. The main environmental problem with fracking is the risk of groundwater contamination. This can happen if the water-chemical mixture or the released gas finds a way through the well into aquifers (and since many aquifers are connected, the contamination is likely to spread), or if the pools containing the contaminated water that flows back out of the well leak or overflow.

Though a newcomer to the energy scene, fracking is already highly controversial, and the water has been muddied by lobbying and polemics. Advocates of shale gas emphasize that it could offer an affordable source of low-carbon energy (at least compared with coal) to reduce dependence on coal and oil. A study conducted in 2012 by the University of Texas found that all stages of the fracking process entailed environmental risks, and that contamination had already occurred in numerous cases (Groat and Grimshaw 2012). A 2010 report by the Worldwatch Institute concluded that hydraulic fracturing does not necessarily entail damage to human health and the environment,

Figure 6.9. A poster announcing a demonstration against fracking in Spain, October 2012. Protests against fracking have accompanied the spread of this controversial extraction method.
Source: Zarateman at Wikimedia Commons.

but that the risks are high and therefore strict regulation is needed (Zoback et al. 2010).

6.4 The Nuclear Tangle

Masataka Shimizu, president of TEPCO, the company that operated the ill-fated Fukushima Daiichi nuclear power plant, was received harshly when he toured a temporary shelter for 1,600 people a few weeks after the 2011 tsunami and nuclear meltdown. The images of executives kneeling and bowing in front of displaced people was broadcast worldwide and became yet another iconic image illustrating the problems with nuclear power.

The prompt evacuation of the area around the reactors limited public exposure to harmful radiation, yet, according to Hirooki Yabe, a neuropsychiatrist at Fukushima Medical University, two years after the twin disaster, "the tsunami-area people seem to be improving; they have more positive attitudes about the future" while "nuclear evacuees are becoming more depressed day by day" (Brumfiel 2013). Uncertainty, isolation, and fears about future health are jeopardizing

Figure 6.10. An anti-nuclear power manifestation in Tokyo six months after the Fukushima disaster.
Source: 保守 at Wikimedia Commons.

the mental health of the 210,000 residents who fled from the nuclear disaster (Brumfiel 2013).

From an environmental point of view, nuclear power is both hero and villain. Because it avoids combustion, no carbon dioxide or air pollutants are emitted. A threefold expansion of nuclear power could contribute significantly to staving off climate change by avoiding 1 to 2 billion tons of carbon emissions annually (Deutch and Moniz 2006). Unfortunately, many of the by-products of uranium fission, including tritium, caesium, krypton, and iodine, are radioactive, and therefore highly dangerous to plant, animal and human health.

Every 12–18 months, nuclear power plants must shut down for a few days for the 'spent' uranium fuel – now radioactive waste – to be replaced. The International Atomic Energy Agency (IAEA) defines radioactive waste as any material that contains a concentration of radionuclides greater than those deemed safe by national authorities, and for which no use is currently available or foreseen.[7] Worldwide

---

[7] A radionuclide is an atom with an unstable nucleus that undergoes radioactive decay, resulting in the emission of gamma rays and subatomic particles. These emissions constitute nuclear radiation.

nuclear power facilities produce each year about 200,000 cubic metres of low- and intermediate-level waste and 10,000 cubic metres of high-level waste.[8]

The big difference between nuclear waste and other forms of pollution is not just the extent of the hazard, but also its life span. Whereas most toxic pollutants are naturally broken down into harmless compounds in a matter of decades, nuclear materials remain highly dangerous for many thousands of years.[9] Such time spans, far longer than the history of human civilisation, create a unique challenge for planning and waste management. Not surprisingly, the IAEA concluded as early as 1989 that "the future of nuclear energy is dependent on our ability to handle and dispose of radioactive waste in a safe and acceptable manner" (Larsson 1989, 25).

Many countries have tried, unsuccessfully, to find permanent sites for nuclear waste. Yucca Mountain, deep in the Nevada Desert, was chosen in 1987 as the answer to the United States' nuclear waste problem, but after just twenty-two years and 9 billion dollars, the project was abandoned (Wald 2009). The same fate has met most proposed long-term disposal projects in the last forty years. Some options such as 'long-term above-ground storage' have not been implemented, some such as 'ocean disposal' or 'Antarctic ice sheet disposal' have been banned by international agreements, and some such as 'outer space disposal', are the stuff of science fiction.

In the absence of permanent solutions, spent fuel is usually stored in steel-reinforced concrete containers next to the nuclear plants. The issue is sensitive, with a stratospherically high NIMBY factor. Some scientists argue that, since radioactive waste held in interim storage gradually decays and becomes easier to deal with, the smartest solution, at least pending a technological breakthrough, is to do nothing (Wald 2009). Opponents counter that this throws the precautionary principle out the window, since radioactive decay takes thousands of years.

---

[8] According to IAEA, there are six general categories of radioactive waste, presented in a decreasing order of risk: (1) spent nuclear fuel from nuclear reactors; (2) waste from the reprocessing of spent nuclear fuel; (3) waste from military programmes; (4) tailings from the mining and milling of uranium ore; (5) waste from hospitals or industry, containing small amounts of mostly short-lived radioactivity; (6) naturally occurring radioactive materials.

[9] Typically, the radioactive life span of a material is measured in 'half-lives'. This refers to the length of time it would take for a material to lose half of its radioactive capacity. Technetium-99, a by-product of uranium fission, has a half-life of 220,000 years, and iodine-129 has a half-life of 17 million years.

6.5 Hydropower: Transforming Entire Ecosystems

Walls of stone will stand upstream to the west
To hold back Wushan's clouds and rain
Till a smooth lake rises in the narrow gorges.
The mountain goddess if she is still there
Will marvel at a world so changed

In the above poem, entitled "Swimming", Mao Zedong expressed his vision of a giant dam on the Yangtze River (Tvedt 2013, 202). This megaproject was not realized in his lifetime, but work began eighteen years after his death, in 1994. When the project was approved, human rights activists and environmentalists expressed concern about a possible social and environmental disaster. To make room for the project, 1.2 million people, living in two cities and 116 towns along the banks of the Yangtze, were forced to relocate with a promise of plots of land and stipends of US$7 a month as compensation.

According to George Davis, a tropical medicine specialist at George Washington University Medical Center, as a result of the dam "there's been a lot less rain, a lot more drought, and the potential for increased disease" (Hvistendahl 2008). In 2008 even the government official in charge of the project admitted that building a massive hydropower dam in an area that is heavily populated, home to threatened animal and plant species, and crossed by geologic fault lines may have been a mistake, as it alters entire ecosystems and may trigger landslides (Hvistendahl 2008). Other major dam projects, such as Egypt's Aswan Dam and India's Sardar Sarovar Dam, also involved huge forced population movements. In most cases, the dislocated populations ended up in slums (McNeill 2000).

There have been several notable dam failures in history. The most catastrophic was the bursting of the Banqiao Dam in China in 1975, killing an estimated 171,000 people and displacing 11 million others. To this day, dam failures are a regular occurrence, though thanks to safety improvements and more accurate geological surveys, human casualties are usually measured in tens rather than thousands.

Today about 800,000 dams operate worldwide, 45,000 of which are taller than 15 metres. The benefits of dams extend beyond the generation of clean electricity. They may also control flooding; their reservoirs provide a reliable supply of water for irrigation, drinking, and recreation; and, by stabilizing river flow, some dams boost navigation and trade (Marks 2007).

Figure 6.11. The Merowe Dam on the Nile in Dar al-Manasir, Nubia, Northern Sudan. Hydroelectric power stations submerge large areas of land because of the requirements of a reservoir. This has major impacts on ecosystems and human communities.
Source: Lubumbashi at Wikimedia Commons.

In the past, hydropower fostered economic and social development in many countries by providing both energy and water management services. However, whereas run-of-river hydro does not significantly alter a river's flow regime, the creation of a reservoir for storage hydropower entails a major environmental change by transforming a fast-running fluvial ecosystem into a still-standing lacustrine one. Through the building of dikes and weirs, hydroelectric structures affect the river's ecology by changing its hydrologic characteristics, and by disrupting the ecological continuity of sediment transport and fish migration. Salmon hatch upstream in a freshwater environment and spend their adult lives at sea. Eel do it the other way around. Dams disrupt the life cycle of both. Nor is such disruption limited to a few species of fish. The construction of the Aswan Dam caused the extinction of the populations of the Nile crocodile downstream and today the papyrus artworks sold to tourists are based on cultivated plantations because the Egyptian wild papyrus has almost disappeared.

Hydropower is approaching its global technical limits (Andrews et al. 2011). Whether it can sustainably contribute to socio-economic development depends largely on how its services and revenues are shared among different stakeholders and how small-scale projects (where there is more room for expansion) can cope with environmental problems.

6.6 Wind: Power Trumps Beauty

In 2007 I conducted an environmental impact assessment of a 20-megawatt wind farm project in central Italy. At the time, wind power was rapidly expanding in Italy, giving rise to a public debate on its merits. I suggested to my client that they pursue a participatory approach, whereby local residents would be informed in a public presentation before the first perimeter fencing went up. The meeting was on a Friday evening and the turnout was high. After the engineers presented the facility design, I spoke of the impacts of wind energy in general and of the specific project. My presentation was interrupted by two elderly gentlemen wearing the traditional "coppola" caps. They stood up and loudly accused me of shilling for a useless project that would poison the air of their beloved mountains. Large-scale wind farms do have serious impacts on the local environment, but air pollution is not one of them. This experience illustrates the often irrational aspects of the NIMBY attitude, and how important it is to consult people, taking the time necessary to dispel unfounded fears and address valid concerns.

In 2012 the property tycoon Donald Trump halted the construction of his luxury golf resort in Aberdeenshire because of a proposed offshore wind farm, and publicly accused Scottish First Minister Alex Salmond of aiming to 'destroy' Scotland (Hubbard 2012). Trump is not alone in objecting to the appearance of wind turbines. Since many of the most suitable locations for wind power are also areas of scenic beauty, wind power tends to collide with local sensitivities and tourism interests. Not surprisingly, a myriad of anti-wind action groups have emerged throughout Europe.

Set against a fiery sunset, the stark silhouettes of wind turbines are difficult to ignore. Some regard them as beautiful because they represent a move towards clean sustainable energy. For others the rotating blades are high-tech intruders upon otherwise pristine landscapes. Whatever its impact on landscape (and this is hard to objectively assess, being largely in the eye of the beholder), the impact of wind power on the way land is used is moderate, as there is nothing to stop a farmer raising animals or cultivating crops among the towers of a wind farm.

The blades of a wind turbine rotate at speeds of up to 200 kilometres per hour and therefore pose a danger to migrating birds, large birds of prey, and bats, which may be struck by the rotor during flight or disturbed if nesting nearby.[10] However, this impact needs to be seen in perspective as there are numerous other ways that humans affect bird populations, many of them far deadlier than wind power. According to one study (Erickson et al. 2005), wind power currently accounts for 0.01 per cent of anthropogenic bird mortality, while domestic cats are responsible for about 10 per cent and windowpanes for more than 55 per cent (see Figure 6.13).

[10] The impact of wind turbines on wildlife, most notably on birds and bats, has been widely studied and documented, although the cumulative impact assessment process is still unsatisfactory with few impact studies considering cumulative impacts. Birds can die in collisions with the moving blades, and by striking towers and power lines. Mortality attributable to collision with turbines varies considerably between different locations, from zero to more than 300 individuals per turbine per year. Wildlife can also be impacted by the siting of turbines, power

(a)

(b)

Figure 6.12. The perception of impacts depends a lot on how data are presented. The image above emphasizes the low impact on land use, while the bottom image shows the visual intrusion of a sprawling wind farm on a natural landscape.
Source: Gian Andrea Pagnoni.

Wind turbines are often installed in agricultural landscapes or on brownfield sites that are ecologically far less sensitive than tropical environments. However, from a conservationist's perspective, the

lines, and access roads. Some species are sensitive to the presence of turbines and can be scared away from their breeding grounds (Benner et al. 1993; de Lucas et al. 2007; Everaert and Kuijken 2007; Janss et al. 2001; Masden et al. 2010, Percival 2007; Sterner et al. 2007; Strickland et al. 2011; Wiser et al. 2011).

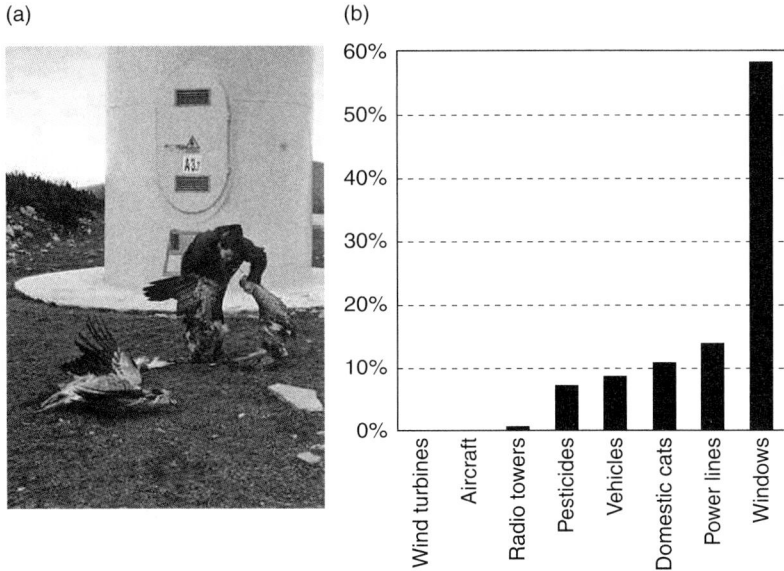

Figure 6.13. Data on the impact of wind turbines on wildlife are often presented in an emotional way. The reality is more complex and varies greatly according to location. The photograph on the left shows two griffon vultures that were struck and killed by turbine blades, on the right a comparison between bird mortality attributable to wind power and other human causes (percent).
Sources: www.iberica2000.org (left) and Erickson et al. (2005) (modified).

death of fifty house sparrows is preferable to that of a single eagle, as the second species is far less common. Yet larks and sparrows tend to feed near the ground well clear of a turbine's rotor, whereas vultures and eagles in search of prey fly at rotor height and pay little attention to the moving blades.

Although a great number of studies have been conducted worldwide on the impacts of wind power on wildlife and ecosystems, the relative biological significance of these impacts remains unclear. The nature of the impact depends hugely on the ecosystem within which a wind farm is placed and on how vulnerable local species are. It is difficult to predict bird and bat fatality rates before a wind farm is built. However, it makes sense to situate wind farms only in areas where birds and bats are scarce, and to use different types and sizes of turbines, depending on the species present (Strickland et al. 2011; Wiser et al. 2011).

Last but not least, we need to mention the noise impact of wind turbines. A 2-megawatt turbine produces 105 decibels of sound at the nacelle. People living 200 metres away would experience constant background noise of 50 decibels, roughly equivalent to a never-ending dinner party on their front lawn. The simple way to solve issues relating to the noise produced by turbines is to install them at a suitable distance from residential areas.

6.7 The Impacts of Solar Technologies

Solar thermal collectors are based on a very simple technology, and contain mostly non-hazardous materials such as copper, aluminium, black paint, plastic, and glass. Their land impact depends on whether they are installed on roofs or on the ground. The only impact of roof-based plants is visual, and even that is minimal; the panels are almost indistinguishable from roof windows, and only the storage tank, a large cylindrical container, is conspicuous. Only in historic city centres can the presence of solar thermal collectors be reasonably impugned, and for that reason they (or just the external storage tank) have been banned from many old town centres.

Unless you have installed solar panels, you are unlikely to spot the difference between a thermal and a PV panel. The tell-tale sign from the ground is the number of panels on the roof. A couple of thermal panels are often enough to meet a household's heating requirements, whereas an entire south-facing roof covered in PV may not meet its electrical needs.

Solar power and heating plants do not produce noise or impact greatly on water supply. Although quite a lot of water is used to manufacture silicon, installed plants consume only the water needed to periodically clean the panels. In dusty and dry locations, periodic cleaning of the panels may be necessary to maintain performance (Arvizu et al. 2011a).

Still under debate is the case of thin-film modules, which usually contain cadmium, a known human carcinogen. BP Solar, a branch of the famous oil company, and the Japanese Matsushita have ended their production of modules containing cadmium, though probably more for reasons of image than because of genuine health risks. Alsema et al. (2006) calculated the amount of cadmium released in the life cycles of various energy sources. They showed that thin-film PV technology releases less cadmium than silicon technology (0.3 grams per GWh instead of 0.9 g/GWh), and much less than the oil industry (43.3 g/GWh) (Sinha et al. 2008). Ultimately, most studies have shown that PV power generation is an environmentally friendly alternative to fossil fuels (Arvizu et al. 2011a).

Some have raised environmental concern about PV once the whole life cycle (from production to disposal) is considered. Indeed, of all industrial processes, silicon production is by far the most energy intensive (it requires about fifty times the energy of steel production). As a result, the indirect impacts of solar PV may be high, mostly through the emission of GHGs during the production phase. Countering end-of-life-cycle concerns is the fact that PV modules are highly recyclable. Silicon does not become less pure over time, but merely needs to be reimpregnated with the 'doping' elements. Risks of contamination can be minimized by take-back programmes that may be paid for by insurance premiums incorporated into the cost of the product (Sinha et al. 2008). Recycling of PV modules is already economically viable, and projections are that 90 per cent of the glass, plastics and metals in PV panels will be recycled in the near future (Arvizu et al. 2011a). The barrier, as is so often with energy, is one of economics. At present, the solar sector is too small in most countries to support a viable recycling sector. However, economics is likely to provide a solution, as the inevitable long-term rise in the global price of fossil fuels will make solar power more competitive and cause the sector to grow.

The economics of concentrated solar power (CSP) require investors to seek out locations with the most concentrated resources (i.e., sunny deserts). Therefore, there is little competition with agriculture or urban development and land impacts are low. Environmental problems mainly arise when protected species of animals and plants are present. At present, CSP plants are only economically viable when located close to large urban areas with an existing power grid. This may change, however, as CSP capacity and economies of scale increase, making it viable to carry power from remote plants. This would open up a range of very attractive sites in many regions of the world, including southern Europe, northern and southern Africa, the Middle East, Central Asia, western China, northwestern India, Australia, Chile, Peru, Mexico, and the southwestern United States (Arvizu et al. 2011a; IEA 2009a).

6.8 Bioenergy: Food Versus Fuel

The Tripa forest in Indonesia's Aceh province was supposed to be a protected area. Yet in 2011 the governor of Aceh granted a permit to the Indonesian palm oil industry to clear more than 1,600 hectares of rainforest for palm oil plantations. Part of the jungle was razed by fire, destroying an ecosystem that had taken centuries to develop and further shrinking the habitat of the Sumatran orang-utan. In September 2012, under pressure from environmental groups, the governor revoked the permit.

The Tripa forest is a small part of a global palm oil–driven deforestation crisis. The palm oil industry is also expanding into wild forests in western and central Africa, threatening important tropical ecosystems (*Scientific American* 2012). However, not all of Africa's forest destruction is driven by powerful industrial interests. A rapidly growing human population is putting intolerable pressure on the continent's forests and wildlife.[11] One of the main causes of deforestation in Africa mirrors the deforestation of Europe a few centuries ago: the demand for charcoal. In Congo, Rwanda, and Uganda, charcoal is not only the primary local fuel for heating and cooking; it is also a valuable means of sterilising water. As the richer countries have seen with the drug trade, it is impossible to stem supply if demand is strong enough. Until the people of Africa's Great Lakes region have access to an affordable alternative to charcoal, illegal felling will continue, and those who get in the way of this lucrative trade, whether a park ranger or one of the last few hundred mountain gorillas, are likely to pay with their lives (Platt 2009).

Bioenergy also has a major impact on GHG emissions. Converting tropical forest or grassland into plantations for biodiesel production results in a net carbon dioxide output 17–420 times greater than the annual GHG emission reductions that these biofuels would provide by displacing fossil fuels. The practice of burning forest to remove trees and increase soil fertility not only reduces the capacity of the land to absorb carbon dioxide but oxidizes the exposed soil, causing massive erosion. The result is a major net increase in carbon emissions (Eisentraut 2010; IEA Bioenergy 2009; Melillo et al. 2009). Petroleum is therefore more sustainable than biofuels produced through forest cleaning. In fact, as a result of extensive deforestation, Indonesia ranks as the world's fourth largest net GHG emitter (Sari et al. 2007). By contrast, biofuels made from wastes or from plants grown on degraded

---

[11] According to John Cleland, an international expert on reproductive issues in Africa, sub-Saharan Africa is "the one remaining region of the world where the population is set to double or treble in the next 40 years" (UNFPA 2011).

Figure 6.14. Deforestation of a tropical forest in Riau province (Sumatra), to make way for an oil palm plantation. When natural ecosystems are turned into energy crop plantations, habitats are lost, often forever. Since natural habitats, by definition, are home to most of the world's plant and animal species, their loss means the loss of biodiversity.
Source: Hayden at Wikimedia Commons.

and abandoned lands can offer immediate and sustained advantages (Fargione et al. 2008).

Bioenergy may lead to greater crop diversity and income opportunities for farmers in wealthier countries, but in many poorer countries where land ownership is more concentrated, it may also cause social problems. According to the World Bank (2008), the cultivation of energy crops in developing countries may add to the problems already caused by the cultivation of cash crops such as cotton and coffee; poor farmers become more dependent on markets that are beyond their control. In such a scenario farmers sell energy crops at an instable price in order to purchase food, also at instable prices. It is often harder to survive on the lowest rung of a cash economy than from subsistence agriculture. Restricted access to land, changing employment trends and the rising cost of food, which is no longer produced directly by the farmer, can create a worse condition than the original one based on traditional subsistence agriculture.

When, in 2009, the European Union made a commitment to meet 10 per cent of its transport energy needs from renewable sources by 2020, it effectively issued a purchase order for millions of tons of food crops. While the Renewables Directive called for biodiesel to be produced within the EU from vegetable oils, bioethanol production would be mostly imported from Brazil, where the associated land impacts would also be felt (Al Riffai 2010). By the end of 2012, biofuels had fallen somewhat out of favour, having been blamed for the global food crises of 2008–2009. In response, the European Union modified its Renewables Directive, providing stimulus funding for the development of second-generation biofuels (made from crop residues rather than food commodities), and limited the use of food-based biofuels from 10 per cent to 5 per cent. The EU Commissioner for Climate Action Connie Hedegaard explained the change of course: "We are of course not closing down first generation biofuels, but we are sending a clear signal that future increases in biofuels must come from advanced biofuels. Everything else will be unsustainable" (EU press release 2012). This change of course on biofuels suggests that a consensus on the extent of the impacts of bioenergy is unlikely to be achieved in the near future.

The expansion of energy crops can also have positive environmental impacts. The air pollution from the combustion of biofuels is considerably lower than from coal or petroleum. Therefore, using biofuels to replace fossil fuels can mitigate climate change, which is likely to be a major driver of habitat loss. Bioenergy production consumes more water than oil production does, yet its impact on water resources is not solely negative. Although energy crops grown conventionally with pesticides and fertilizers threaten water supplies through chemical run-off and eutrophication (organic pollution of aquatic ecosystems), perennial woody crops such as palm oil consume far less water than annual crops such as wheat (Chum et al. 2011).

All in all it is very difficult to calculate whether a practice, commodity, or product is sustainable. To do so, we must conduct what is called a life cycle assessment, taking into account its environmental and social impact from production through manufacturing to disposal.

6.9 The Impacts of Geothermal Energy

The hot water tapped from the Earth at geothermal plants is not pure. It contains minerals that leached from the reservoir rock, variable quantities of GHGs (mainly carbon dioxide), and some hydrogen sulphide, which is responsible for the rotten-egg smell that often occurs near vents and geysers. The chemicals present in geothermal fluids, if released at the surface, can have adverse effects on the ecology of rivers, lakes, or marine environments. In the past, before the dangers were sufficiently understood, geothermal fluids were often released at the surface. Nowadays, it is usually injected back into the reservoir after passing through a steam turbine. The twin advantage of this approach is that it helps to maintain high pressure in the reservoir and avoids contamination of watercourses.

There is no combustion involved in geothermal energy. In closed-loop power plants, where the geothermal fluid passes through a heat exchanger before being reinjected into the ground, carbon dioxide emissions are close to zero. In direct heating applications, emissions are also negligible.[12] Therefore, geothermal energy does not contribute to global warming (Goldstein et al. 2011).

---

[12] In Reykjavik, Iceland, the CO_2 content of thermal groundwater used for district heating (0.05 milligrams of CO_2 per kilowatt-hour of heat) is lower than that of the cold groundwater. In China (Beijing, Tianjin, and Xianyang) it is less than 1 gram of CO_2 per kilowatt-hour of heat (Goldstein et al. 2011).

Piercing the Earth's crust to depths of hundreds or thousands of metres affects its geological structure. As with other deep drilling projects, pressure or temperature changes induced by stimulation, production, or injection of fluids can lead to geo-mechanical stress changes. The rock interstices, there for millions of years, are quickly emptied of hot liquid and filled with colder liquid coming from the power plant. Some claim that geothermal energy production may cause micro-earthquakes, steam eruptions, and ground subsidence. Yet, no buildings have been significantly damaged by shallow earthquakes resulting from the extraction and reinjection of geothermal fluid. Ground vibrations or noise from EGS demonstration projects have not been large enough to lead to human injury or significant property damage. At a few high-temperature geothermal fields, the reduction in underground pressure because of extraction of geothermal fluid has led to ground subsidence. Fluid reinjection helps to maintain pressures in the reservoir and avoid subsidence.

Geothermal energy belongs in the lowest category of land use impact. However, one of the biggest constraints on geothermal energy is the issue of landscape impact. The insulated steel pipelines that carry steam and hot water to a power plant from drilling points have a major visual impact on the landscape. They also act as barriers to grazing animals and tractors. There are, however, several successful examples of how geothermal energy can be reconciled with tourism in places of great scenic beauty, including Wairakei in New Zealand and the Blue Lagoon in Iceland (see Figure 6.15) (Goldstein et al. 2011).

6.10 The Impacts of Ocean Energy

The physical presence of new structures may affect marine ecosystems, both above and below the surface. Some fish, such as salmon and eel, migrate between fresh and salt waters to spawn, and marine energy devices may act as a physical barrier. Rotors can also injure or kill fish or dolphin in a similar manner to wind turbines striking birds and bats, although the blades of water turbines, for current or tidal energy, move at much slower speeds than wind turbines do, and thus the likelihood of blade strike is lower.

Like conventional power plants, which use natural watercourses as a coolant, ocean thermal energy conversion (OTEC) devices exchange large volumes of deep and shallow water to take advantage of the difference in temperature. If pursued on a large scale, this practice would negatively affect the marine plants and animals in both

Figure 6.15. These two geothermal energy plants produce visible impacts on the landscape, though hikers are still free to roam at Larderello (Italy), and the buildings in a formerly pristine landscape at the Blue Lagoon (Iceland) do not seem to bother the bathers.
Sources: above, Gian Andrea Pagnoni; below, Mark Hintsa at Flickr.

strata. Moreover, large pipes would run along the ocean floor, altering the environment for planktonic organisms at the bottom of the marine food chain (Boehlert and Gill 2010; Lewis et al. 2011).

However, in 2001, the British Government concluded that "the adverse environmental impact of wave and tidal energy devices is minimal and far less than that of nearly any other source of energy, but further research is required to establish the effect of real installations" (House of Commons 2001). Marine renewable energy has great promise but, unlike wind, sun, and biomass, it is still a long way from commercial viability. The environmental impacts of ocean energy are difficult to quantify since very little research has been conducted to date.

6.11 Taking Carbon out of the Equation

The goals of the Kyoto Protocol have not been reached; far from reducing carbon dioxide emissions by 5.2 per cent from the 22.7-billion-tonne level in 1990, we managed to increase them to 33 billion tonnes by 2010. Moreover, there is no end in sight, as current projections indicate that, for the foreseeable future, fossil fuels will continue to be burned in quantities incompatible with reducing or even stabilizing GHG concentrations (see Section 5.5).

That is not to say that nothing has changed in the way we use fossil fuels. In the last twenty years, major technological advances have been made in reducing air pollution. The most important of these are the catalytic converters fitted to car exhausts and the scrubbers and filters added to the smokestacks of power plants. It is likely that future advances will further increase our ability to reduce the air pollution impacts of fossil fuels.[13] However, there is no way to filter out the carbon dioxide emitted from the burning of fossil fuels. The only possible way to reduce emissions without reducing the use of combustion fuels would be to capture the carbon dioxide as it is released and store it deep underground to prevent it entering the atmosphere.[14]

Capturing the carbon, at least from power plants, would not be difficult. The major problem arises with transport and storage. This is not a problem of technology (for decades oil companies have been injecting carbon dioxide into oil fields to push the remaining oil to the surface) but of cost and logistics. Unlike nuclear waste, which is highly concentrated and could be stored in a few underground locations, carbon capture and storage (CCS) would require vast underground spaces.

---

[13] It should be noted, however, that these technologies are expensive and are not used in all countries. Most of the new coal-fired power stations built in the last twenty years have been in emerging economies such as India and China, where environmental standards are lower and oversight less strict than in the wealthier Western countries. For similar reasons, pollution from vehicle fumes tends to be higher in these countries.

[14] Suggested sites include depleted oil and gas fields and salt mines. Initially, there were also suggestions that carbon dioxide could be pumped into the deep oceans, but this idea was abandoned when it became clear that this would increase the problem of ocean acidification.

Even if such locations were readily available, long-term predictions about underground storage security are uncertain, and carbon dioxide might, one day, leak into the atmosphere. Injecting millions of tons of carbon dioxide per year into the ground also poses potential risks to soil and groundwater. Capturing and storing carbon would also be very costly in energy terms and could lead to a doubling of plant costs, in which case the current cost advantage of fossil fuels over renewables would be nullified (Haszeldine 2009).

To date, there are only a few pilot-scale CCS facilities in operation. Indeed, at the seventh Carbon Capture & Sequestration conference in 2008, Greenpeace launched a critical report on CCS, portraying it as little more than a smokescreen to rehabilitate the dirtiest fossil fuel under the banner of 'clean coal' technology, and saying its capacity to reduce carbon dioxide emissions is minimal (Rochon et al. 2008). Others are more upbeat. According to the IEA, "Carbon capture and storage is more than a strategy for 'clean coal'. CCS technology must also be adopted by biomass and gas power plants; in the fuel transformation and gas processing sectors; and in emissions-intensive industrial sectors like cement, iron and steel, chemicals, and pulp and paper" (IEA 2009c, 4). The IEA believes that CCS could play an important role in the global transition to a sustainable low carbon economy, accounting for up to one-fifth of global GHG emissions reductions from the energy sector by 2050 (IEA 2009b). However, even proponents of CCS admit that this solution will not be available at industrial scale before 2030 (Haszeldine 2009; Rochon et al. 2008; Smil 2010).

6.12 The Monetary Costs of Energy Production

Many people believe that the path to renewable energy has been blocked by powerful lobbies with vested interests in maintaining the fossil fuel economy. This is true up to a point, insofar as human beings are apt to place personal interests above the common good. But there are other reasons for the continued dominance of fossil fuels. The technology for producing fossil fuels is well established, and a huge global infrastructure exists to deliver them to their end users. Replacing this will require significant investment.

In most countries, a litre of petrol (that is, mineral oil extracted from the earth, refined, and delivered to the pump) costs less than a litre of soda (that is, water extracted from the earth, enriched with sugars and flavours, and delivered to the store). This illustrates the point that, in purely monetary terms, fossil fuels are very cheap. So far, only a few renewable energy sources (hydro, geothermal, and biomass) are already competitive with coal, oil and gas.

The costs of a litre of petrol, a kilogramme of coal, a cubic metre of gas, or a kilowatt-hour of conventional electricity are calculated based on the market price of the primary resource, the capital invested to build the various facilities and infrastructures required to produce the secondary resource and deliver it to the consumer, their operating and maintenance (O&M) costs, and the cost of decommissioning those facilities at the end of their life. The cost breakdown for renewables is somewhat different, since, apart from bioenergy, the primary resource is free. However, because the technologies of renewable energy are relatively immature, the initial investment and O&M costs are high.

Investment costs are sensitive to several variables. Materials and technology change and sometimes improve rapidly, making a previously unaffordable process commercially viable. Steel prices affect the initial cost of any facility, while solar PV is highly sensitive to fluctuations in silicon and aluminium costs. Economies of scale also play an important role, as has been seen in the case of wind and solar where mass production significantly reduced the cost of the turbines and the PV cells.

Notwithstanding these variables, energy economists are able to calculate the relative costs per kilowatt-hour of different energy sources. These are referred to as the levelized cost of energy (LCOE) (see Figure 6.16).[15]

On average, the costs of renewable energy are higher than those of fossil-based energy. Indeed, only large-scale hydropower, geothermal projects of more than 30 megawatts and (if the cost of carbon is reflected in the markets) wind onshore power plants are already competitive in the marketplace without incentives. Renewable energy may be competitive in niche power markets, such as remote areas where no grid-based electricity is available. Some applications may also be competitive in certain contexts, such as solar thermal in sunny regions, biofuels in Brazil and pellets in Europe (Fischedick et al. 2011).

---

[15] According to the IEA, the LCOE is defined as "the ratio of total lifetime expenses versus total expected outputs, expressed in terms of the present value equivalent" (2005, 174). To calculate LCOE we divide the costs of setting up and running an energy facility (investments, O&M, fuel, and decommissioning costs) by its lifetime energy output. LCOE does not take into account subsidies and incentives. LCOE calculations also vary from one site to another, depending on fuel prices, resource availability, capacity factors, interest rates on investment, and O&M costs.

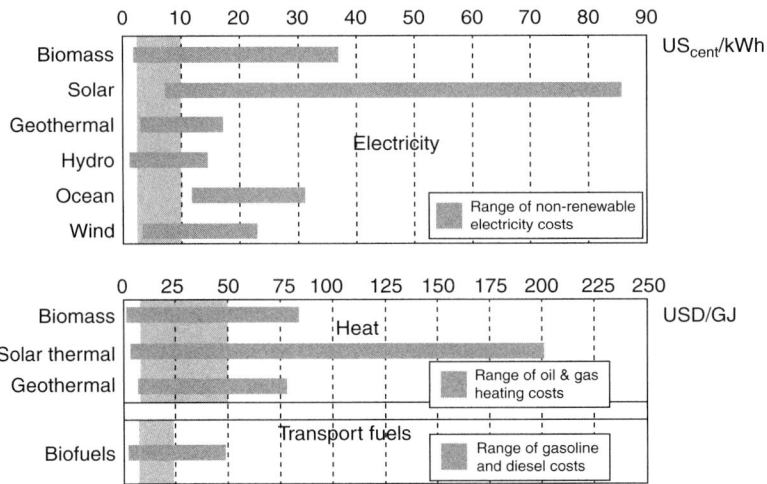

Figure 6.16. LCOE for selected renewable energy technologies compared with nonrenewable sources. This shows that renewables technologies are cost-competitive with conventional sources only in specific circumstances.
Source: Fischedick et al. (2011) (modified).

However, the monetary costs of a litre of petrol or a kilowatt-hour of fossil-generated electricity do not tell the full story.

6.13 External Costs of Energy Production

When the Nobel Prize–winning economist Milton Friedman was asked to sum up his ideas in one sentence, he responded, "There ain't no such thing as a free lunch."[16] The free lunch was a marketing ploy used by American bars in the nineteenth century. Patrons were offered free food if they purchased just one beer. However, the food was heavy on salt, and as a result the customers purchased more beer to quench their thirst. The analogy to energy economics is that no energy comes to us for free; if the cost is not obvious, it is hidden. In the case of sun, water, wind and geothermal energy, the primary source is free, but the costs of harnessing it are high. In the case of fossil fuels, the costs of extracting the primary source are low, but there are major social

---

[16] Final thoughts for students at the twenty-fifth annual Young America's Foundation National Conservative Student Conference. See http://www.youtube.com/watch?v=WC0elAPyXhU.

and environmental costs involved in burning them. The hidden costs of fossil fuels include air and water pollution which affect human, animal, and plant health. To these, we can add future problems associated with climate change, such as desertification, drought, flooding, sanitation, and agricultural productivity. It is difficult to put a monetary price on these impacts as they manifest themselves slowly over a long period of time and can arise far from the pollution source in completely different ecosystems and societies.

Typical external costs include emissions from the burning of fossil fuels and the emissions from the industries that manufacture materials for the energy facility, such as concrete, steel, silicon, or plastics. To understand the full impact of energy use on GHG concentration, the entire life cycle of a particular energy technology (from mining the raw materials to decommissioning the facility) must be taken into account (see Figure 6.17).

As we have seen, renewable technologies can have negative impacts on water, land use, soil, ecosystems, and biodiversity. However, they produce negligible emissions of GHGs and other air pollutants. Therefore, their impact on climate change and the associated external costs are usually low (see Figure 6.18).[17] Considering the whole life cycle of different technologies and including all externalized costs, if renewables were to replace fossil fuels, the overall economic, social, and environmental costs of energy would decrease and the net impact would be positive (Fischedick et al. 2011).

---

[17] The relationships between exposure and health impacts are estimated on the basis of epidemiological studies, which show a reduction of quality and duration of life. The concept of "value of life years lost" is used to assess the external costs of increased mortality and is expressed as money per ton of carbon or CO_2 released or per kilowatt-hour produced. A number of studies have provided values of US$17–350 per tonne of CO_2. As a result of the many possible approaches to quantifying damage, the range of the estimated social costs of carbon cover three orders of magnitude. Many of the studies published on the social costs of carbon have been criticized for this reason. However, a value of about US$17 per ton of CO_2 released into the atmosphere can be given as a benchmark lower limit. For other hazardous chemicals, the figures are US$4,000–10,000 per tonne of sulphur dioxide, US$2,000–10,000 per tonne of nitrous oxides, and US$10,000–30,000 per tonne of particulates PM 2.5 (Fischedick et al. 2011). All U.S. dollar figures are based on the 2005 value.

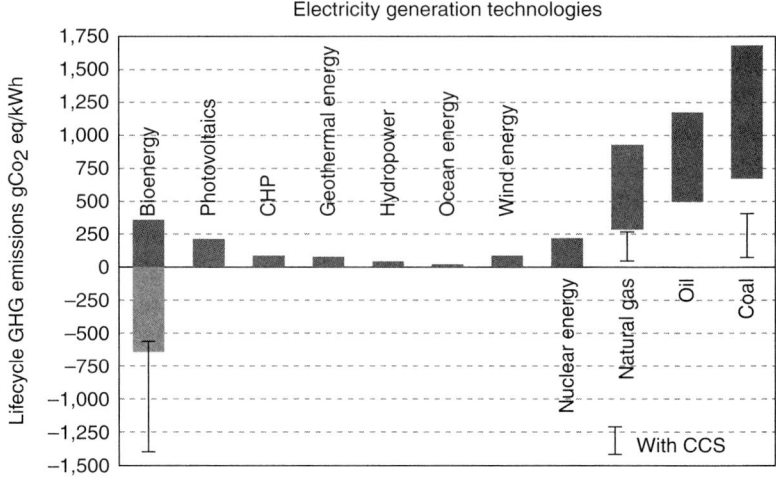

Figure 6.17. GHG emissions (in grams of CO_2 equivalent per kilowatt-hour of energy produced) of different electricity-generating technologies. Negative figures show avoided emissions. If CCS is used, the ranges are lower. For biopower, avoided emissions are higher than emissions estimates. All technologies entail some emissions (at the very least during the construction phase), but only bioenergy offers the possibility to remove CO_2 from the atmosphere.
Source: Sathaye et al. (2011) (modified).

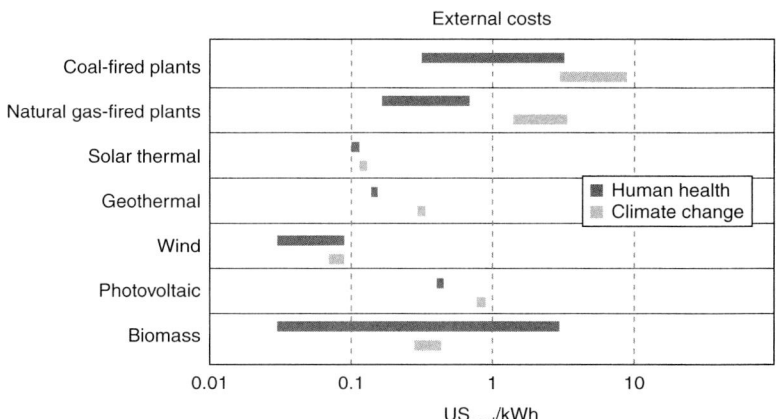

Figure 6.18. External cost ranges, in terms of human health (air pollution) and climate change (GHG emissions), of renewable and conventional energy sources. External costs attributable to climate change dominate in fossil energy if not equipped with CCS.
Source: Fischedick et al. (2011) (modified).

Table 6.2. External costs (U.S. cents per kilowatt-hour) of renewable and conventional electricity production based on Central European conditions

	PV 2000	PV 2030	Hydro	Wind Onshore	Wind Offshore	Geo-thermal	Solar Thermal	Lignite	Coal	Natural Gas
Climate change	0.86	0.48	0.11	0.09	0.08	0.33	0.11	9.3	7.4	3.4
Health	0.43	0.25	0.075	0.09	0.04	0.15	0.11	0.63	0.46	0.21
Material damages	0.011	0.008	0.001	0.001	0.001	0.004	0.002	0.019	0.016	0.006
Agricultural losses	0.006	0.004	0.001	0.002	0.0005	0.002	0.001	0.013	0.011	0.005
Sum	1.3	0.74	0.19	0.18	0.12	0.49	0.22	9.9	7.9	3.6

Source: Fischedick et al. (2011) (modified).

6.14 Buying and Selling Pollution

One way to deal with the external costs of energy production is to impose a tax on the particular activity that produces those costs. Notable precedents are taxes on cigarettes and alcohol, well known to affect health. A more complex approach involves trying to trade pollution. 'Emissions trading,' or 'cap-and-trade', was one of the cornerstones of the 1997 Kyoto Treaty. It is a market-based mechanism to control air pollution by providing economic incentives to reduce emissions of pollutants. A central authority (usually a government or intergovernmental authority) sets a limit or cap on the amount of a pollutant that may be emitted (e.g., one million tonnes of CO_2 per year). This cap is allocated or sold to firms in the form of emissions permits which represent the right to emit or discharge a specific volume of the pollutant.

The rationale behind this approach is that, since most manufacturing and conventional energy industries cannot avoid pollution, this should be recognised by granting permits equivalent to the emissions they expect to release in a given period of time. Since there is a maximum cap imposed by the government, the total number of permits cannot exceed the cap. If a company exceeds its allotted quota of pollution, it must purchase additional credits on an open market. Those companies who invest in filters and cleaner production methods can then sell their 'spare' permits to the polluters. Faced with the choice of investing in the technology to filter out air pollutants or paying for additional emissions permits, power utilities and other polluters would make the more economical choice.

Emissions trading forces the buyer to pay for polluting, while the seller is rewarded for having reduced emissions. There are currently several emissions trading programmes in operation worldwide, covering several different pollutants. The largest is the European Union Emission Trading Scheme, whose purpose is to reduce GHG emissions. In the United States there is also a national market in sulphur dioxide emissions to reduce acid rain and several regional markets in nitrogen oxides.[18] Regional initiatives have also been launched in Canada,

---

[18] A recent EPA (United States) analysis claims that the Acid Rain Program, by reducing sulfur dioxide and nitrogen oxide emissions, will cut premature human deaths due to fine particle pollution by between 20,000 and 50,000 per year, and between 430 and 2,000 premature deaths attributable to ground-level ozone (smog). The Nitrogen Oxides Budget Trading Program saved an estimated 580 to 1,800 American lives in 2008. The Acid Rain Program has reduced sulphur dioxide

Australia, and New Zealand, and the city of Tokyo has implemented its own trading scheme. Some companies have even implemented internal emissions trading schemes. BP, for example, managed to reduce its greenhouse gas emissions by one-tenth between 2000 and 2010, a level of success that most national governments have yet to emulate (Victor and House 2006).

The greatest difficulty in limiting carbon dioxide emissions is the fact that international agreement is required. Even a gradually phased-in system of taxes, fines, and permits would only work if implemented worldwide. If only some countries impose carbon taxes, industry is likely to migrate to those countries where the burden is lowest. The people who negotiated the Kyoto treaty foresaw this problem, and created a mechanism whereby countries can gain 'credits', known as Certified Emission Reduction (CER) units, for steps taken to reduce GHG emissions, such as capturing methane, and trade these in exchange for credits allowing them to emit more carbon dioxide. This 'Clean Development Mechanism' is designed to allow countries maximum flexibility in balancing the needs of development with their commitments to reduce emissions. Credits may be traded with other countries or used domestically to increase the allowable quota of GHG emissions. The Clean Development Mechanism allows industrialized countries to buy CER units where they are cheapest globally, thus providing an incentive to support emission reduction projects in developing countries.

Conventional utilities and even oil and gas companies are already starting to invest in renewable energy. If emissions trading eventually catches on at a global level, the price of carbon will rise, and so too will the viability of energy sources that produce few or no emissions. In the short term, this may favour nuclear energy, since it already has the economy of scale to compete with fossil fuels, but in the longer term the shift is likely to be towards sustainable renewable energy.

6.15 The Energy Cost of Energy Production

A sweet-toothed child drinking soda through a straw will persevere until every last drop of soda has been drawn from the bottom of the bottle. Leaving aside the impolite noises that ensue, the child is getting

emissions in the United States from 17.3 million tons in 1980 to about 7.6 million tons in 2008. Emissions of nitrogen oxides decreased by 43 per cent between 2003 and 2008. See http://www.epa.gov/captrade/.

a very low return on energy investment for these final drops. In the early days of oil, when the first Texas gushers were struck, it took the energy of one barrel of oil to extract 100 barrels more (Hall and Klitgaard 2012). By stark contrast, the ratio for deepwater drilling in the Gulf of Mexico today is approximately one in ten (Moerschbaecher and Day 2011). This ratio is known as energy return on investment (EROI).

In the 1980s Charles Hall, an ecologist at the State University of New York, went down to the coast of North Carolina, looking for a place where fish migration could be studied from an energetic perspective. He measured the ecosystem productivity (energy availability through food) of the freshwater environment and found some very clear patterns: "The fish would migrate to capitalize on the abundance of energy for the first years of life, and then the young fish would migrate downstream into a more stable but less productive environment. The study found that fish populations that migrated would return at least four calories for every calorie they invested in the process of migration" (Inman 2013b). With this study, Hall was the first to describe and apply the concept of EROI, which has since been applied to more complex energetic systems.

The EROI of a particular energy source is one of the most valuable indicators of its viability. As we drill deeper and deeper to reach unexploited oil reserves, we become more like the child slurping through the straw. For example, the steps involved in mining and processing tar sands to produce oil are highly energy-intensive. First, the tar-like bitumen must be separated from the heavy soil. The bitumen is then heated to convert it into crude oil. Further energy is required to transport it by pipeline and tanker to refineries, to turn the crude into gasoline or other fuels, and for the final cleanup and land reclamation. However, the rising investment required to recover conventional crude oil (and its correspondingly decreasing EROI) is turning unconventional sources of oil and gas into more attractive energy sources (Dale et al. 2011; Inman 2013a).

There is great divergence between the EROI of different energy sources and for the same energy source over time. Because of their low EROI, some renewables are viable only with subsidies. At the opposite end of the spectrum, hydropower delivers an EROI of as much as 260 to 1, making it the most energy-effective of all (see Table 6.3). The reason for this is that most of the work in a hydropower plant is done by the sun through evaporation and run-off, and the technology involved is both well established and cost effective.

Table 6.3. *EROI of energy commodities from various sources*

Process	EROI
Oil and gas (1930)	> 100
Oil and gas (1970)	30
Oil and gas (today)	11–18
Coal (1950)	100
Coal (today)	18–80
Natural gas	7
Bitumen from tar sands	2–4
Shale oil	5
Hydrogen	0.67–2.5
Ethanol from sugarcane	0.8–10
Ethanol from corn	0.8–1.6
Biodiesel	1.3–5.5
Electricity from coal	14–43
Electricity from coal after CCS	4–24
Electricity from natural gas	6–26
Electricity from nuclear fuels	4–68*
Electricity from hydropower	50–260
Electricity from photovoltaics	4–12
Electricity from CSP	4–10
Electricity from wind	10–25
Electricity from geothermal	11
Heat from solar thermal (warm water)	30–50
Heat from solar thermal (space heating)	5–10

* Risk costs not taken into account.
Sources: Dale et al. (2011), Hacatoglu et al. (2012), Inman (2013a), Kubiszewski et al. (2010), Kümmel (2011), Murphy and Hall (2010), Raugei et al. (2012), Smil (2008).

Coal, at the point it is removed from the ground, has a high EROI, but this falls sharply when it is burned to generate electricity, and further still if the carbon dioxide is captured and stored. The EROI of nuclear energy depends a lot on the technology used. Older reactors are far less efficient and require more highly enriched fuels. Moreover, the EROI of nuclear power generally does not include the costs of decommissioning the plant, of waste disposal over thousands of years, or the potential for accidents. Solar PV has a low EROI mainly because silicon (at present the main material in commercial solar cells) is highly energy-intensive to manufacture. As technology improves and greater economies of scale develop, the EROI of solar PV is certain to increase. Indeed, this is true of all the alternative technologies, while the EROI of fossil fuels is destined to continue falling (Dale et al. 2011).

The current energy economy requires that fuels have an EROI above five. Only hydro, fossil fuels, and solar thermal are well above this benchmark. As easily accessible oil reserves dwindle, the EROI of fossil fuels will decrease, and societies will spend more on energy production. (Inman 2013a, 2013b).

EROI does not take into account all the benefits and drawbacks of a fuel. It does not consider the environmental costs or the quality of the source. Nevertheless, EROI offers a very useful way of assessing the advantages and disadvantages of a given fuel or energy source. It allows investment to be concentrated on the most economically viable sources, as market prices are likely to reflect EROI (Hall and Klitgaard 2012; Inman 2013a; King and Hall 2011).

6.16 Heading towards a Sustainable Energy Supply

According to Prince Hassan bin Talal of Jordan, chairman of the governing board of the Arab Thought Forum, "More than 40 years ago the Apollo Space Program was launched to fulfil the old dream of taking man into outer space. Today, we have a bigger dream, to restore balance between man and his home planet, Earth" (DESERTEC 2009, 8).

According to the ecologist Charles Hall and the economist Kent Klitgaard, in terms of spending power, middle-class incomes in the United States have not increased in the last twenty years. In fact, without massive borrowing, both from its own citizens and foreign investors in the form of government bonds, the U.S. economy would have shrunk rather than grown in the last ten years. The same is true of many Western countries, especially the stricken economies of southern Europe. Klitgaard and Hall put this down to politicians' and economists' tendency to focus exclusively on growth. Economics seems to be about money, but fundamentally it's about stuff. Indeed, what money gives us is access to food, a roof over our heads, petrol for our car, and the car itself. All that stuff is extracted from the Earth's ecosystems, and these, contrary to the working assumptions of most economists, are not infinite (Hall and Klitgaard 2012; Inman 2013b).

According to the Millennium Ecosystem Assessment (MEA), a series of reports on the state of the Earth commissioned by the United

Nations and involving more than 1,000 of the world's leading scientists, "Over the past 50 years, humans have changed ecosystems more rapidly and extensively than in any comparable period of time in human history, largely to meet rapidly growing demands for food, fresh water, timber, fiber and fuel. This has resulted in a substantial and largely irreversible loss in the diversity of life on Earth [...]. The changes that have been made to ecosystems have contributed to substantial net gains in human well-being and economic development, but these gains have been achieved at growing costs in the form of the degradation of many ecosystem services [...]. These problems, unless addressed, will substantially diminish the benefits that future generations obtain from ecosystems" (2005, 1).

In the coming decades we are likely to see an increase in global energy consumption. Even with improved technology and efficiency, we cannot escape the fact that energy production always entails some kind of environmental trade-off. We may aspire to a future in which that trade-off does not involve wars, poisoned rivers, or melting ice caps, but even then some sacrifice will be required, such as an eagle sliced in two by the blades of a wind turbine, the sight of hillsides covered with PV panels, or villages relocated for a hydropower dam.

We must find and embrace sustainable ways to improve quality of life on a limited planet. This will not happen by the action of governments alone. As individuals, we must all take some responsibility for reducing waste and producing energy in a more sustainable way. This is likely to involve a combination of global and local action, such as agreeing to tackle climate change at a global level and powering our local communities from our backyards.

7
Energy from My Backyard

7.1 The Insatiable Demand for Energy

Around the turn of the nineteenth century, the human population of this planet topped one billion for the first time. A century later, it had increased by roughly 600,000 people. Two centuries later (in October 2011), it already reached 7 billion. Clearly, something extraordinary has happened in the last hundred years. Despite two world wars and numerous smaller conflicts that cumulatively claimed 231 million lives during the twentieth century (Leitenberg 2006), population has grown exponentially. So too has human demand for the finite resources of the global ecosystem. In the last forty years, world population has doubled, the global economy has grown fifteen times, the number of cars sixteen times, and fertilizer use sixfold. This has resulted in shrinking forests and more land for food,[1] vast mines defacing landscapes, and vastly increased water consumption (Biello 2011b).

Primary energy production provides a useful indicator of the impact of human population growth and economic development on the Earth's resources. Energy production increased more than tenfold during the last century, and since the first oil crisis in 1973 it has doubled from 255 to more than 530 exajoules per year (IEA 2012a). Global electricity generation currently stands at roughly 190 exajoules per year (38 per cent of primary production). Most of this energy (62 per cent) is lost in conversion as heat.

Coal is the dominant primary source for power production, accounting for 40 per cent of all electricity generated. Oil is of minor

---

[1] More acreage was converted to cultivation of crops between 1950 and 1980, than from 1700 to 1850 (Biello 2011b).

Table 7.1. *Changes in energy production compared with population increase during the last three centuries*

Year	1700	1800	1900	2000
Fossil fuels and electricity (EJ/year)	0	1.5	22.5	390
Biomass (EJ/year)	10	12	21	42
Population (billion)	0.63	0.98	1.65	6.09

Source: Smil (2008).

importance for electricity generation (6 per cent) but dominates the transport sector. A considerable amount of oil is also used to produce products such as plastics, lubricants, chemicals, and fertilizers. Natural gas, currently at 21 per cent, is rapidly increasing its share in power production and is likely to overtake coal in the next decade. Renewables also play a significant role (19 per cent), the most important by far being hydroelectricity (16 per cent). Despite its abiding unpopularity, nuclear power retains an important share (13 per cent) of the global power mix. 'New renewables', although rapidly expanding, still account for only 3 per cent of global electricity generation.

The most energy-hungry sector of human activity is industry, which consumes 105 exajoules per year, though transport (93 EJ per year) and domestic consumption (85 EJ per year) are not far behind. Someone living in a highly developed country may be surprised to read that the largest domestic energy sources are biomass and (mostly agricultural) waste, attributable to massive consumption by billions of people in poorer countries (IEA 2012a and IEA website).

Though global primary production has doubled since the early 1970s, the energy mix has not changed much. Only nuclear power, with its sixfold growth, has significantly increased its share (see Figure 7.1). Despite the introduction of nuclear energy and various new renewables, fossil fuels still account for 85 per cent of global primary energy, down just a single percentage point from 1973.

However, public perception of fossil fuels has changed considerably during that time. They began to fall out of favour during the first oil crisis in 1973, as people became more aware that the oil supply was finite. The Gulf War of 1991 and Iraq War of 2003–2011 underscored the geopolitical dangers of relying on a resource that is concentrated in one region of the world. Ultimately, the strongest argument against fossil fuels may be environmental, as carbon-heavy fuels are known to be the main cause of climate change.

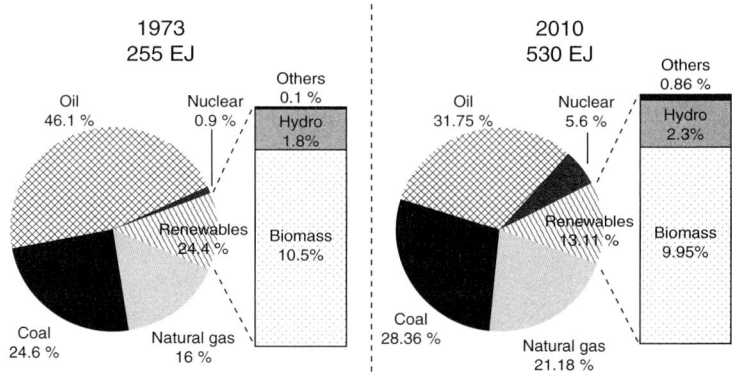

Figure 7.1. Since the first oil crisis, total primary energy supply has doubled, yet the energy mix has not changed considerably, and fossil fuels still dominate. New renewables (represented in the diagram above as 'Others', which include wind, solar, geothermal and ocean energy, and exclude biomass and hydropower) still account for less than 1 per cent of primary energy supply.

Sources: IEA (2012a) and IEA website.

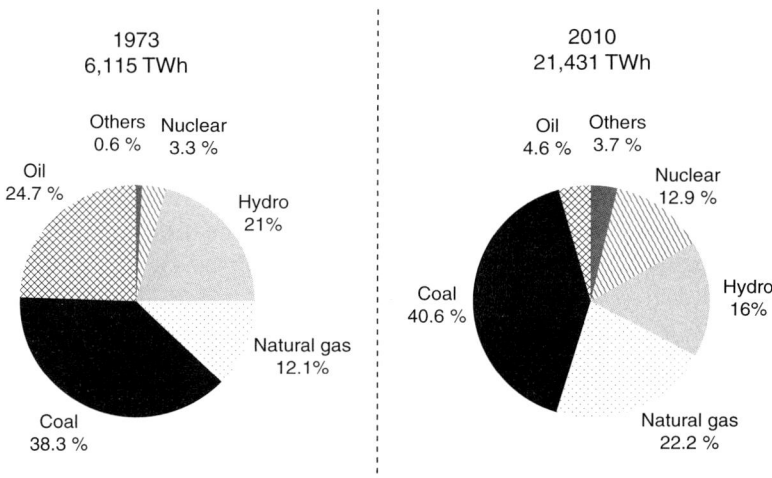

Figure 7.2. Electricity generation has tripled since 1973. Diversification is more evident in this sector than in primary energy supply. A drastic reduction in the use of oil for power generation has facilitated the emergence of nuclear, natural gas and new renewables. Coal remains the most important fuel for power generation.

Sources: IEA (2012a) and IEA website.

Despite changes in attitudes to fossil fuels, we have yet to implement a viable alternative, and so modern human societies stumble along in a confused way, still wedded to fossil fuels but no longer happy with the relationship. In fact, far from serving divorce papers, we have recently renewed our vows, through the rapid development of unconventional sources, such as tar sands and shale gas.

Contrary to popular belief, America's main foreign source of oil is not in the Middle East but immediately to the north. More than a quarter of U.S. oil imports come from Canada, half of it in the form of bitumen melted out of the Alberta oil sands (Biello 2012b). The discovery and use of hydraulic fracturing ('fracking') has brought about a similar revolution in the gas industry; shale gas now accounts for 25 per cent of the American gas supply, up from just 1 per cent in 2000 (Yergin 2011).

7.2 The Stillborn Renaissance of Nuclear Energy

One of the most important developments of the twentieth century was the replacement of coal by oil as the world's major energy source. The British industrialist William Armstrong predicted in 1863 that England would cease to produce coal within two centuries because it "was used wastefully and extravagantly in all its applications" (Higgins 2007). Armstrong was a strong advocate of solar power and hydroelectricity because "the solar heat operating on one acre in the tropics would … exert the amazing power of 4,000 horses acting for nearly nine hours every day" (Higgins 2007). Although hydropower played an important role during the early stages of industrialization in many countries, recent global development has been mainly based on the exploitation of fossil fuels (Sathaye et al. 2011).

In 1973 the so-called traditional renewables (biomass, waste, and hydroelectric power) accounted for roughly 12 per cent of global primary energy production, compared with a whopping 86 per cent produced by fossil fuels. The share provided by 'new' renewables (wind, solar and geothermal) was insignificant (0.1 per cent). At that time the fastest-growing alternative to fossil fuels was nuclear energy, and many people expected nuclear power to gradually replace fossil fuels.

Two of the great advantages of nuclear power are the maturity of the technology and the widespread occurrence of uranium ore. Compared with other non-fossil energy sources such as wind and solar power, nuclear power was quick to reach maturity; just sixteen years separated the first laboratory chain reaction (1942) from the first commercial power plant (1958). Yet it was not until the early 1970s, as the oil market teetered, that nuclear power went mainstream. Between 1970 and 1980 global nuclear capacity grew from about 10 to 350 gigawatts. Thirty years later, in 2010, that capacity remained stagnant at 375 gigawatts. Despite rapid development and generous government subsidies, nuclear power failed to live up to its promise as a major alternative to fossil fuels. Energy scientist Vaclav Smil describes nuclear power as a "successful failure"; a technological success but an economic failure (2010, 42). The reasons for this failure are complex and varied, but they can be boiled down to two factors: risk and cost (Sections 5.7 and 6.5).

The recent emergence of power-hungry economies in Asia and South America seemed to augur, at first, a revival of nuclear power. As concerns about carbon dioxide emissions grew, nuclear power re-emerged as a kind of 'light-green' energy source. Several leading environmentalists, including former Greenpeace director Stephen Tindale, began to lobby in favour of nuclear power. According to Tindale, "nuclear power is not ideal but it's better than climate change" (Connor 2009). The accident at the Fukushima Daiichi plant in 2011 put an abrupt stop to the incipient renaissance and rehabilitation of nuclear power. Its echo has gone far beyond Japan, reigniting the almost-dormant debate on the safety of nuclear energy in many countries.

7.3 Future Energy Use in a Business-as-Usual Scenario

Many agencies, including the IEA, BP, the United Nations, and Greenpeace, publish regular energy outlooks and scenarios for further discussion among citizens and governments. Energy scenarios provide descriptions of how alternative energy conditions could emerge in the foreseeable future. They analyse available data on social, economic, and environmental issues, take into account potential future technological breakthroughs and political turmoil, explore plausible cause-and-effect links, and highlight key decisions and their likely consequences. More than 150 energy scenarios are currently available (Fischedick et al. 2011). They range from studies that assume a 'business as usual' approach, where current energy policies continue more or less unchanged, to more optimistic projections, which assume that action is taken to reduce greenhouse gas (GHG) emissions and achieve sustainability.

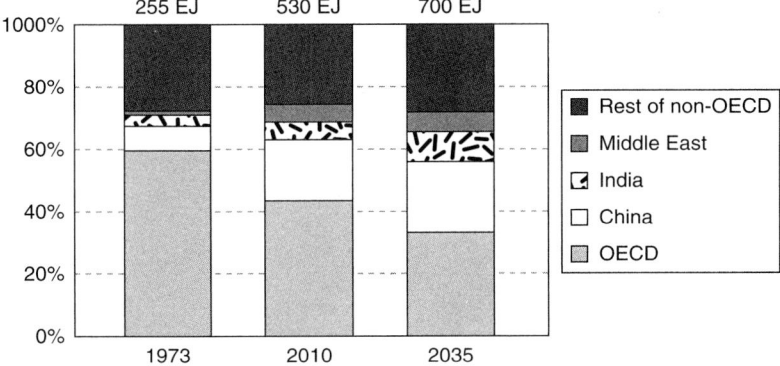

Figure 7.3. Share of global energy demand. In the IEA baseline scenario, global energy demand rises by more than one-third in the period to 2035, driven by rising living standards in China, India and the Middle East.
Source: IEA (2012a) (modified).

All of these scenarios envisage a significant growth of total primary energy production throughout the first half of the twenty-first century. According to the IEA, primary energy production will rise from its current 530 exajoules to between 600 and 1,000 exajoules by 2050 (Chum et al. 2011). Where the scenarios differ most is in their predictions of which energy sources will predominate (Section 7.6).

In the IEA 'Current Policies' scenario, which assumes no major change in energy economics, demand in rich countries will rise only slightly between now and 2035, while the 'emerging' economies of Asia and the Middle East will drive global energy markets.[2]

Major shifts in the global energy economy are expected as unconventional fossil resources (mainly the Canadian tar sands and the U.S. shale gas) are unlocked. Natural gas will overtake coal as the largest primary energy source and North America will become a net oil exporter by 2030. As the United States becomes self-sufficient in fossil fuels, the shift of oil trading towards Asian markets will accelerate. By 2035, almost 90 per cent of the oil resources drilled in the Middle East will be sold to Asia (IEA 2012a).

---

[2] According to the IEA baseline scenario, OECD energy demand in 2035 will be just 3 per cent higher than today, while the share of non-OECD countries will rise from 55 per cent in 2010 to 65 per cent in 2035 (IEA 2012a).

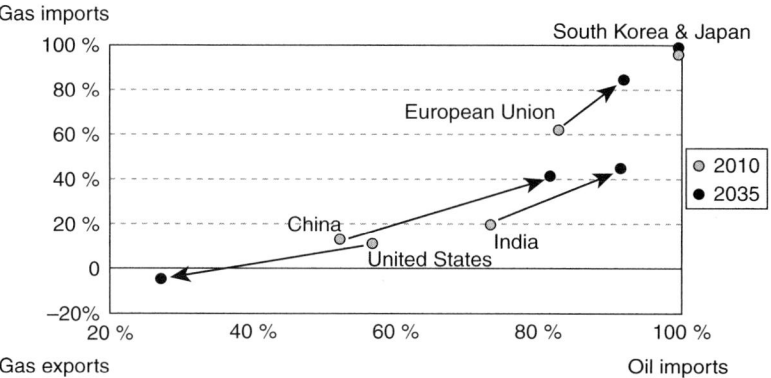

Figure 7.4. Net oil and gas import dependency in selected countries according to the IEA baseline scenario. While dependency is expected to rise in many countries, the United States will move to the opposite direction.
Source: IEA (2012a) (modified).

Oil production will grow, but less rapidly than other sources, and nuclear power will maintain its current share of electricity generation. Renewables, driven by incentives, falling unit costs, and rising fossil fuel prices, will have the highest growth rate, and even in the sober IEA baseline scenario they represent half of the projected 5,890 gigawatts of global power capacity added by 2035 (see Figure 7.5).[3] However, government subsidies for renewable energy will continue to be dwarfed by those for fossil fuels (in 2011 global fossil fuels subsidies amounted to $US523 billion, six times more than all subsidies to renewables). This policy will continue to distort energy markets, and although the share of fossil fuels in the global energy mix is projected to fall from 85 per cent to 75 per cent by 2035, they will remain dominant (IEA 2012a; Figure 7.1).

Recent economic development has not been evenly spread. Nearly a billion people are undernourished, while 222 million metric tons of food are wasted every year in the wealthier countries. The world's richest 500 million people produce 50 per cent of the world's

---

[3] According to the IEA baseline scenario, the share of renewables in the primary energy mix is expected to rise from the current 13 per cent to 25 per cent by 2035. Most of this additional renewable capacity will come from biomass (20 per cent), followed by wind (2.7 per cent) and solar (1.4 per cent). In electricity generation, the share of renewables is projected to grow from the current 20 per cent to 31 per cent, catching up with coal (IEA 2012a).

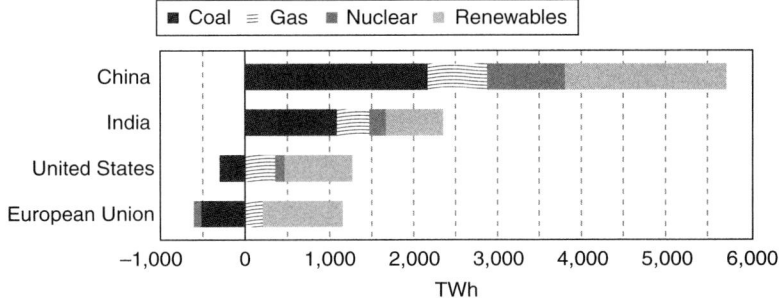

Figure 7.5. Change in electricity generation by 2035 according to the IEA baseline scenario. The demand for electricity in emerging economies will drive a 70 per cent increase in worldwide production. Use of coal is expected to increase in Asia and decrease in the United States and Europe. Growth of nuclear power will mostly occur in China. Gas and renewables will increase worldwide, the latter accounting for half of new global capacity.
Source: IEA (2012a) (modified).

carbon dioxide emissions, whereas the poorest half a billion are responsible for just 7 per cent (Biello 2011b; FAO 2009).

Increases in energy production will have a major impact on the world's supply of fresh water. Water consumption is projected to rise by 85 per cent between 2010 and 2035 as a result of more water-intensive power generation and growing reliance on biofuels and unconventional fossil fuels (IEA 2012a). According to Alexander Mueller, assistant director-general for environment and natural resources of the United Nations Food and Agriculture Organization (FAO), "It is time to stop treating food, water and energy as separate issues and tackle the challenge of intelligently balancing the needs of these three sectors, building on synergies, finding opportunities to reduce waste and identifying ways that water can be shared and reused, rather than competed for." (FAO 2011b)

Since the problem of unsustainable growth became apparent in the early 1970s, an enormous body of evidence has been gathered to try to understand its extent. According to the United Nations Millennium Ecosystem Assessment (MEA), approximately 60 per cent of what it calls "ecosystem services" (the resources provided to humans by ecosystems, including fresh water, fisheries, air and water purification, and the regulation of climate, natural hazards and pests) are being degraded or used unsustainably (MEA 2005).

The Worldwide Fund for Nature (WWF), one of the most prominent environmental NGOs, developed its own system for measuring the impact of human activity on the biosphere: the ecological footprint. This measures humanity's impact on the Earth in terms of the amount of biologically productive land and sea required to deliver the resources we use and to absorb our waste. In 2005, the global ecological footprint was 17.5 billion global hectares, which equates to 2.7 hectares per person.[4] On the supply side, WWF estimated the total productive area or biocapacity of the Earth at just 13.6 billion global hectares, or 2.1 global hectares per person (WWF 2008, 15). This effectively means that we are already living beyond our biological means. Every industrialized country, including China, has exceeded its permissible footprint, and several have exceeded it richly. The UK average ecological footprint was 6.3 global hectares, and the city of London draws upon a biocapacity 293 times its area (Blewitt 2008).

One issue looms very large in all discussion of future energy use: climate change. What began in the late 1960s as the predictions of a small group of scientists has grown into an avalanche of data that has proven beyond reasonable doubt that the rapidly growing concentrations of GHGs in the Earth's atmosphere are the result of human activity.[5] Measured concentrations of carbon dioxide, the most prevalent GHG, have increased from 280 to more than 390 parts per million in the last 150 years. As a consequence, the global average temperature has increased by 0.76 degrees Celsius. If we follow the "business as usual" approach, global average temperature is expected to rise during this century by between 1.1 degrees Celsius and 6.4 degrees Celsius (IEA 2012a; Moomaw et al. 2011).

More than twenty years after the signing of the United Nations Framework Convention on Climate Change (UNFCCC), the dangers of climate change are clearer than ever. Many of the more than 150

---

[4] The global hectare is a unit that measures the average biocapacity of all biologically productive areas on Earth in a given year. These include cropland, forests, and fisheries, but do not include deserts, glaciers and the open ocean. The 'global hectare per person' refers to the ability to produce resources and absorb wastes per person.

[5] The global increases in carbon dioxide concentration are attributable primarily to fossil fuel use and land use change, while those of methane and nitrous oxide are primarily attributable to agriculture. The current shares of GHG emissions from different sources are: 56.6 per cent CO_2 from fossil fuels, 17.3 per cent CO_2 from deforestation and decay of biomass, and 2.8 per cent CO_2 from other sources, 14.3% CH_4, 7.9% N_2O and 1.1% fluorinated gases (Moomaw et al. 2011).

scenarios currently available take into account climate policies, yet there is still a lack of concerted global action. The position of the largest emitters at the international climate change conferences that have been held every year since 1995 might be summed up as: "we recognize the need to jump, we may eventually jump, but we won't jump first." This has created a kind of limbo between acceptance of the problem and the willingness to act. Currently, only the countries of the European Union and Australia have accepted binding emission reduction targets.

The closest the global community has come to concerted action on climate change were the Cancún Agreements of 2010, in which all parties to the UNFCCC agreed that they should reduce their carbon dioxide emissions to avoid a rise in global average temperature exceeding 2 degrees Celsius above pre-industrial levels. However, this represented only agreement on what needs to be done, not a binding commitment to act. To remain below or at the 2-degree threshold, atmospheric GHG concentrations must not rise above 450 parts per million of carbon dioxide equivalent.[6] This means that global emissions must start to decrease, instead of continuing to increase, no later than 2015 and must fall by 50–85 per cent by 2050 (IPCC 2007). A future world in which this target is achieved is referred to as the '450 Scenario' (IEA 2012a).[7]

7.4 A Brighter Vision: Imagining What's Possible

Analysts at agencies such as the IEA are paid to be hard-nosed. Their job is to advise governments on the probable future of energy use. The trouble with focusing on what is probable rather than what is possible is that one automatically tends towards a conservative perspective and an assumption that change will be slow and incremental. Such analysis excludes the visionary. Yet, who but a visionary could have predicted many of the technological and social changes that have taken place in the last century?

Thomas Watson, then chairman of IBM, famously predicted in 1943, "I think there is a world market for maybe five computers"

---

[6] 'Carbon dioxide equivalent' (CO_2-eq) expresses the impact of different GHGs in terms of the amount of CO_2 that would cause the same amount of warming.
[7] The 450 scenario is a policy aiming to provide a 50 per cent chance of capping CO_2 at 450 parts per million and consequently limiting global increase in temperature to 2°C more than the average in preindustrial times (IEA 2012a).

(Swann 2009, 141). Nearly half a century earlier, Mark Twain predicted the invention of a device eerily similar to the Internet.[8] How could the head of a computer company get it so wrong, and a writer, without any technical expertise, come so close to foreseeing the most important technical innovation of the modern age? The reason lies in the differing perspectives of the two men. Whereas Mr. Watson was focused on making products that would sell well during his tenure as chairman, Mr. Twain had the luxury of allowing his imagination free rein. If we are to extricate ourselves from our current dependence on carbon-based energy, and develop societies and economies that are sustainable, the ability, and indeed the audacity, to explore the outer reaches of the possible rather than the near shores of the probable will be needed.

The largest but least tapped form of energy on Earth is solar radiation. According to Gerhard Knies of the DESERTEC Foundation, installing solar technology in a small fraction of the world's deserts (an area roughly the size of Japan) would suffice to meet global energy supply. DESERTEC is a highly ambitious and visionary project that has the support of some of Europe's biggest power utilities (e.g., E.ON, RWE, and Abengoa). Its goal is to meet a substantial part of the energy needs of the Middle East and North Africa, and about 10 per cent of Europe's electricity demand, by 2030 (see Figure 7.6). The abundance and intensity of sunlight in the Middle East and North Africa, resulting in higher per unit yield than is possible in Europe, would compensate for the additional transmission costs and electricity losses through long-distance cables. By transforming some of the Earth's least productive land into a precious energy reservoir, DESERTEC has the potential to boost sustainable energy supply while side-stepping social and environmental objections (Andrews et al. 2011; DESERTEC 2009; IEA 2010b).

DESERTEC, though still embryonic, is an inspiring example of what could be achieved with regional cooperation. But it does not deliver a model of sustainable energy supply that can be applied globally. That task was recently taken up by Mark Jacobson, a professor of civil engineering at Stanford University, and Mark Delucchi, a researcher in transportation studies at the University of California. In

---

[8] In a short story called "From the *London Times* of 1904", which he published in 1898, Twain describes an invention called the telelectroscope which connects to the phone system: "The improved 'limitless-distance' telephone was presently introduced, and the daily doings of the globe made visible to everybody, and audibly discussable too, by witnesses separated by any number of leagues" (Ketterer 1984, 128).

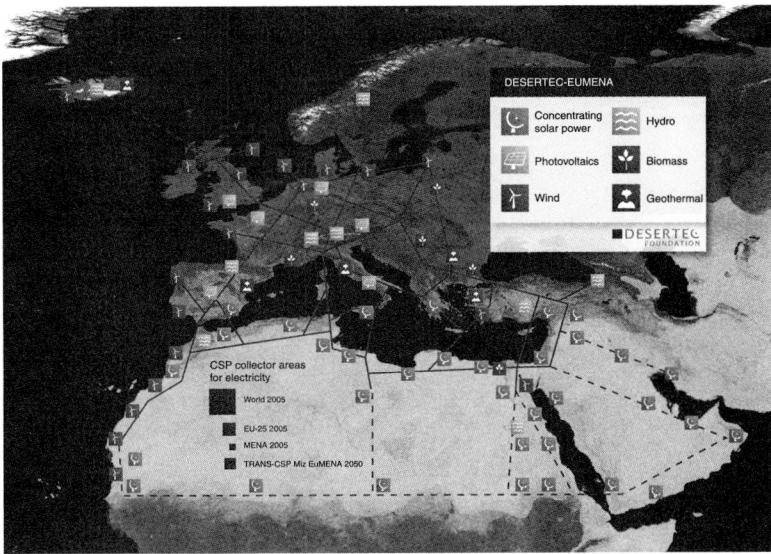

Figure 7.6. The DESERTEC project to supply solar power to the Middle East and North Africa and part of Europe by 2050. The red squares indicate the land area covered by CSP plants that would be needed to meet global electricity consumption (17,000 TWh per year), the consumption by the member states of the European Union (3,200 TWh per year) and by the countries in the Middle East and North Africa (600 TWh per year). The square marked "TRANS-CSP Mix 2050" indicates the area needed to supply energy for seawater desalination and two-thirds of electricity consumption in the Middle East and North Africa and about one-fifth of European electricity consumption (2,940 TWh per year) in 2050.
Source: DESERTEC (2009).

2011 they published a rigorous plan for supplying all energy for all purposes (electric power, transportation, heating, and cooling) from a mixture of renewable sources (Jacobson and Delucchi 2011a, 2011b). They excluded from their model all forms of combustion, even biomass, focusing instead only on wind, sun, and water.

They propose the following mix of technologies deployed in suitable locations: 51 per cent of global demand from wind thanks to 3.8 million 5-megawatt turbines; 40 per cent from the sun, through 89,000 photovoltaic (PV) and CSP plants and 1.7 billion rooftop PV systems; and 9 per cent from tidal, geothermal and hydroelectric plants. According to the authors, only 0.59 per cent of the world's land surface

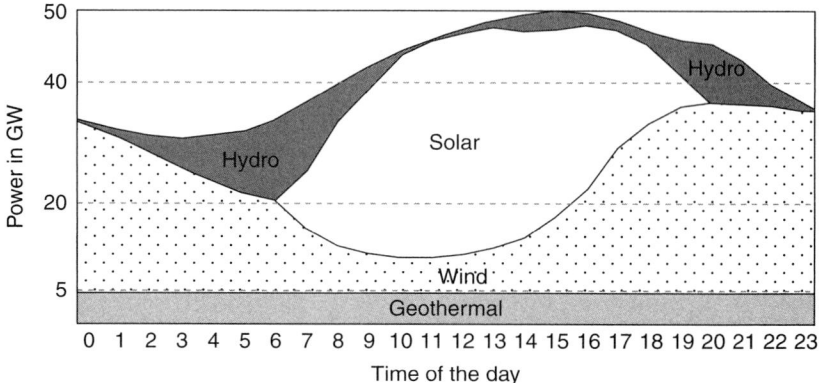

Figure 7.7. Because wind and solar power are intermittent, Jacobson and Delucchi considered ways to ensure steady power supply using different sources at different times of day and night. This diagram projects California's mid-summer electricity demand satisfied by renewables only in 2020. Geothermal provides base-load supply (instead of fossil fuels), wind and solar work alternately at night and during the day, and peak-load is met by hydropower.
Sources: Jacobson and Delucchi (2009, 2011a, 2011b).

would be required to fulfil this model. To ensure a reliable energy supply everywhere, such a system would need to connect resources across long distances through the construction of long grids, and to store energy produced in excess at given times (e.g., in both windy and sunny periods).

According to the authors, a lot can also be achieved by increasing the efficiency of energy use. They estimate that, by improving technologies and reducing the distance between the sites of production and consumption (though small-scale renewables), world power demand can be reduced by 30 per cent. In Jacobson and Delucchi's model, electricity will become the central commodity in the energy economy. The transport sector will abandon oil in favour of hydrogen fuel and electric vehicles. Excess power will either be stored or used to produce hydrogen.

Such scenarios promise a bright future for the power utilities. However, a lot of things need to happen before they can be put into practice. With the exception of hydroelectric plants, current renewable capacity is just a tiny fraction of what would be required under

Jacobson and Delucchi's plan (see Section 7.1). Moreover, we know that, in addition to technology, market forces come into play and, of the renewables, only hydro and geothermal are currently competitive with fossil fuels. However, the authors believe that a steady rise in fossil fuel prices as reserves dwindle will serve to make wind, solar and water power competitive with coal, oil and gas. They project that, by 2030, all new energy demand could be met by renewables, with full replacement by 2050.

The two researchers explain that "building such an extensive infrastructure will take time. But so did the current power plant network. And […] if we stick with fossil fuels, demand by 2030 will rise to 16.9 TW, requiring about 13,000 large new coal plants, which themselves would occupy a lot more land, as would the mining to supply them" (Jacobson and Delucchi 2009, 61). Hence, the barriers to this virtuous plan are not technological or economical, but primarily social and political (Jacobson and Delucchi 2011a, 2011b).

7.5 Reality Check: How Renewables Fare in the Current Market

In 1979, at the height of the world's second major oil crisis, U.S. President Jimmy Carter held a press conference on the roof of the White House. Standing next to a newly installed solar water heater, Carter announced that the United States would invest a billion dollars over the following twelve months to kick-start an energy revolution. He hoped that, by the year 2000, the United States would obtain 20 per cent of its energy from the sun. Carter conceded that this goal was an ambitious one, and predicted that "a generation from now" the White House solar heater would either be "a small part of one of the greatest and most exciting adventures ever undertaken by the American people, or a museum piece" (Yergin 2011, 523). It followed the latter route, as just seven years later, under President Reagan, the solar heater was removed from the roof. Today it is a permanent exhibit at the Carter Presidential Library, and with a 0.1 per cent share of global primary energy production, the solar revolution is still waiting to happen.

In 1979 the United States was the leading force in the nascent solar industry. At the turn of the twenty-first century, another country took the lead, and for the same reason that the United States lost it: political will. In the midst of the political turmoil and jubilation that accompanied German reunification in 1990, a law was quietly passed that would remake Germany's renewable energy sector. The 1991 Feed-in Tariff Law, the last piece of legislation passed by the West German parliament, required power utilities to buy electricity from renewable producers at rates well above market price. The costs would then be spread across the system. Germany was not the first country to devise such a system of market subsidies – the United States had again led the way more than a decade earlier with PURPA,[9] a federal law that requires major power utilities to buy electricity from smaller producers at above-market prices. The difference between the United States and Germany on this point was that over the next twenty years successive German governments maintained and expanded the system of feed-in tariffs, while in the United States the federal supports for renewable energy were undermined, leaving a few states such as California to pursue a more progressive path on their own.

This example illustrates the importance of political will for the emergence of energy alternatives. The model presented by Jacobson and Delucchi, while certainly ambitious and visionary, is not technically implausible. Most energy scenarios (whether technical or political) assume that the share of renewables in the global energy mix will rise significantly in the next twenty years, and renewables are expected to become the dominant low-carbon energy option by 2050 in the majority of currently available scenarios (see Section 7.6). However, whether we see a slow incremental emergence, such as that predicted by the IEA, or the revolution advocated by Jacobson and Delucchi, will depend more on resolve than on technology.

All the energy technologies that have emerged in the past two centuries have been boosted by either market pressure or state support. In the case of the coal-powered steam engines that drove the Industrial Revolution, the profits from mechanised production offered ample incentive for technological innovation. The market for the first commercially available petroleum product – kerosene – was driven by the rapid decline of whale populations. Yet subsidies have also been used to encourage the development of new energy sources. The oil and gas industries have long benefited from subsidies and tax relief as incentives to exploration. In the early twentieth century, many

---

[9] The Public Utility Regulatory Policies Act (PURPA) was passed in 1978 by the United States Congress as part of the National Energy Act.

governments invested heavily in large-scale hydroelectric plants. After the Second World War, solar power emerged from the American space programme and nuclear power from the deep coffers of military research. The latter has been particularly generously funded. In the period from 1945 to 2007, the U.S. nuclear industry received more than 50 per cent of all federal research funds (Pfund and Healey 2011).

The energy landscape of the United States is currently in flux thanks to a sudden abundance of cheap natural gas from the hydraulic fracturing of shale ('fracking'). Energy experts refer to this as the 'shale gale'. It has resulted in a sudden drop in funding for alternative sources, whether from government or venture capitalists. With the U.S. government under intense political pressure to cut spending and investors more interested in quarterly earnings than long-term prospects, the temptation of short-term gain is hard to resist. The words of oil tycoon and erstwhile champion of wind power T. Boone Pickens sum up the prevailing rationale: "Today we have the gas. We're fools if we don't use it" (Biello 2013e). For the moment, fracking is uncommon outside North America. Some countries, such as France and Germany, have even banned it because of environmental concerns.

The 'shale gale' illustrates the structural difficulty of extricating our societies from fossil fuel dependency. Governments, private investors, and the general public tend to embrace renewables in much the same way that yo-yo dieters adopt exercise regimes: enough to see some results, then it all seems like too much trouble and they fall back into the old habits. The analogy is apt, as both cases refer to energy use and overuse, and in both cases systemic change is needed to alter the existing pattern of behaviour.

The task of changing the fossil-fuel-based energy system is hardly less onerous than that of an obese person changing their eating and exercise habits. In fact, in the case of the energy economy, there is an additional challenge: infrastructure. Coal mines, oil and gas fields, refineries, pipelines, tankers, power plants, transformers and transmission lines, filling stations and hundreds of millions of engines that run on petroleum derivatives comprise the most elaborate, extensive and expensive energy infrastructure humanity has ever created. According to Smil, replacing the current supersystem "with an equally extensive and reliable alternative based on renewable energy flows is a task that will require decades of expensive commitment. It is the work of generations of engineers" (2012).

Economics will play a crucial role in the emergence of renewables. For most renewable technologies, the unfortunate truth is that they cannot compete with conventional energy sources in the current marketplace without state subsidies. There are three main reasons for this: energy return on investment (EROI), capacity and infrastructure.

The EROI of a given energy source is a good indicator of its economic viability (see Section 6.15). Modern economies require energy sources with an EROI of five or above (i.e., each unit of energy invested will yield five or more units). As things currently stand, the EROIs of all renewable sources except hydropower and solar thermal lag behind those of fossil fuels. However, the EROI of fossil fuels is likely to rise as easily accessible oil and gas fields dwindle, and utilities are forced to pay the costs of reducing pollution (Inman 2013a, 2013b).

Most renewable technologies have a low capacity factor. This is because the energy sources they rely on are intermittent. Nuclear power plants are the most efficient with capacity factors above 90 per cent. By stark contrast, a PV plant installed in central Spain, the sunniest place in Europe, barely reaches 20 per cent capacity factor. A well-located inshore wind turbine may achieve capacity of 25 per cent to 30 per cent, or 40 per cent offshore (see Section 4.1, Table 4.1).

The issue of low capacity is also linked to the need for new power infrastructure. To guarantee a steady supply of electricity, renewable plants need to be widely dispersed on a continental scale. This would require a significant extension of the power grid, an expensive and environmentally challenging undertaking that would encounter many NIMBY objections. The current grids are mostly unsuited to large-scale use of renewables. They were designed to carry power from a centralized power station to millions of diffuse consumers. The scenario proposed by Jacobson and Delucchi would turn this model around: instead, power would be supplied by many decentralised small plants. This would entail a complete rethinking of the grid technology and many additional high-voltage transmission lines to link production and consumption sites.

Germany has an average insolation of roughly 1,000 kilowatt-hours per square metre per year, lower than southern European countries such as Spain, Italy, and Greece (1,400–1,900 kilowatt-hour per square metre per year) and much lower than Arizona, Egypt, or central Australia (2,000–2,300 kilowatt-hour per square metre per year). So why has Germany taken the lead in solar PV over the last two decades? According to Smil, the answer is simple: "It happened for the best reason there is in politics: money. Welcome to the world of new renewable energies, where the subsidies rule and consumers pay" (2012).

Feed-in Tariffs

The electricity generated from PV plants is currently about five times more expensive than that from conventional sources (coal and natural gas). Were it not for economic incentives, most forms of renewable energy would not be competitive with fossil fuels. These incentives take a number of different forms, and the different approaches reflect economic and financial priorities. Asia has preferred the growth of the domestic technology sector and has subsidized the manufacturers through tax breaks and other incentives, thereby reducing the market price of the technology. In Europe and North America, the priority has been to kick-start a transition from fossil fuels to alternative energy sources and the preferred method has been to offer state-guaranteed above-market prices for the energy sold. These state subsidies are known as feed-in tariffs. They allow eligible renewable electricity generators (whether a family with solar panels on the roof of their home or a major company operating offshore wind farms) to be paid over a fixed long-term period (typically twenty years) a guaranteed premium price for any renewable electricity they produce and feed into the grid.

(continued)

> The payment is usually administered by the utility company or grid operator and is derived from an additional charge for electricity (or heat) that is imposed on national or regional customers. In this way, the added cost of the renewable energy is spread around to all consumers, whether they chose renewable energy or not. Tariffs may be differentiated by technology type, size, and location. For example, if a country is concerned about landscape impacts and prefers to support energy integration within energy-efficient urban areas, the government may provide higher tariffs for small roof-mounted PV plants and lower tariffs for large biomass-fuelled power plants.

In 2004, at a time when the price of electricity in the United States and Germany was about €0.06 per kilowatt-hour and €0.20 per kilowatt-hour, respectively, the feed-in tariff on PV installations offered by the German government guaranteed investors as much as €0.57 per kilowatt-hour for twenty years. With such generous subsidies, it is no wonder that Germany (followed by Spain and Italy) quickly saw a boom in solar PV. A similar phenomenon occurred in the United States with bioethanol and in China with wind power. Benefits have been so generous that in recent years China and the U.S. have complained about each other on a few occasions to the World Trade Organization for unfair market policies (Doom and Goossens 2012; Palmer 2010).

Thanks to subsidies, the renewable sector is growing almost everywhere. The pace of growth is so fast that the IEA recently updated its projections. The 2009 IEA baseline scenario predicted that the share of renewables in the global primary energy mix would grow modestly from 13 per cent to 14 per cent by 2030.[10] Just three years later, the IEA revised its forecast, predicting a great leap forward, and a 25 per cent share by 2030. Not all observers are convinced that the current growth of renewables is sustainable. According to Smil, "Such high growth rates are typical of systems in early stages of development, particularly when the growth has been driven primarily by subsidies. Projections of wind-power generation into the future have been misleadingly optimistic, because they are all based on initial increases from a minuscule base" (2012).

---

[10] The baseline scenario describes a future in which governments are assumed to make no changes to their existing energy policies and measure. The virtue of this scenario is that it provides a picture of how global energy markets may evolve if the underlying trends in energy demand and supply do not change (IEA 2012a).

A Smarter Grid

In August 2003, several power lines in northern Ohio brushed against overgrown trees and shut down. The alarm software failed, leaving local operators unaware of the problem. Transmission lines surrounding the failure spot, already fully loaded, were forced to bear more than their safe quota of electricity. A power plant automatically shut down in response, destabilising the system's equilibrium. More lines and more plants dropped out. The cascade continued, faster than operators could control, and within eight minutes the largest blackout in North American history occurred, leaving 50 million people across eight states and two Canadian provinces without power. In the following two months, major blackouts also occurred in the United Kingdom, Denmark, Sweden and Italy (Massoud and Schewe 2008).

Today's power grid evolved in the early 1900s when localised grids were constructed and eventually connected to each other. The idea of having a national electricity grid was to allow regions with surplus capacity to export their power to regions with a deficit. By the 1960s, the electric grids of developed countries had become very large and highly interconnected, reaching the overwhelming majority of the population. In acknowledgement of its ability to improve people's lives by transforming diverse sources of primary energy into a clean, controllable energy carrier capable of being transmitted over long distances, the power grid was voted "the greatest engineering achievement of the 20th Century" by the American National Academy of Engineering (NAE 2000).

Power grids, as they are currently conceived, deliver power in one direction, from centralized power stations via high-capacity power lines branched to supply industrial and domestic users. Power stations are strategically located close to reserves or supply lines (mines, wells, railways, rivers or ports). However, as consumption grows and energy supply diversifies, more sophisticated control systems become necessary. Electrical engineers foresee a future in which the power grid is integrated with information technologies and the Internet. Such a 'smart grid' would ensure a more reliable supply of electricity, reduce vulnerability to natural disasters or terrorist attacks, and facilitate a switch from the traditional unidirectional supply based on large power stations to the two-way flow of intermittent, decentralized and small-scale renewable power production (Behr and ClimateWire 2011).

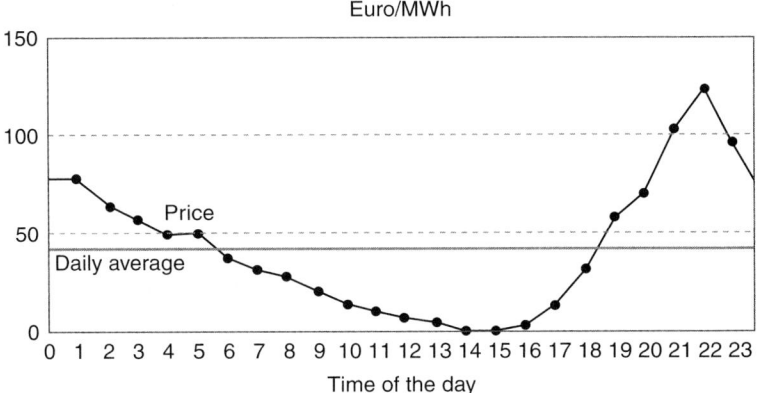

Figure 7.8. The Italian electricity market on June 16, 2013. Thanks to concomitant high levels of solar radiation and wind, renewables satisfied the entire national electricity demand, allowing market prices to fall to zero for a couple of hours.
Source: GME (Italian National Manager of Energy Markets).

On June 16, 2013, in Italy, for two hours in the early afternoon the price of electricity fell to zero (QualEnergia 2013a).[11] This was because in those two hours solar, wind, and hydropower satisfied 100 per cent of the Italian electricity demand. This had a knock-on effect in neighbouring France, which, in recent decades, has exported a lot of power to Italy. Most electricity in France is nuclear. However, nuclear power plants are inflexible – running full-blast all the time – and cannot be ramped up or down in response to sudden fluctuations in demand. Therefore, the day after the Italians experienced a surge in renewable energy, the French felt a painful slump in electricity export prices (Morris 2013).

This was a very dramatic, and short-lived, example of the effect renewables are already having in several energy markets. According to Alessandro Marangoni, chief executive officer of the independent energy consultancy Althesys, "this single event cannot be taken as a reference for a careful management of the electricity market, which

---

[11] Because electricity is difficult and costly to store, prices fluctuate far more than for other commodities, and negative prices are even possible (where suppliers pay customers to take their excess power). As the renewables sector has expanded rapidly in Germany and Italy, driven by generous state subsidies, power utilities have found that they need to adjust their supply chains to avoid generating more power than they can sell.

needs, now more than ever, a new market design in light of a profound change in the structure of supply" (QualEnergia 2013b). A study by Althesys showed that the price in the Italian power market spike no longer coincides with midday peak demand for electricity (see Figure 7.8).[12]

7.6 The Crystal Ball: Future Scenarios for Energy Use

It is difficult to predict the future of renewable energy because so many different factors are involved. However, we can be pretty sure of two things: over the next 20 years, environmental concerns will become more pressing and most renewables will not be able to compete with fossil fuels without subsidies. For those two reasons, neither the IEA baseline nor the most sustainable scenarios are likely to be realised in the near future. Fischedick et al. (2011) recently provided a broad scientific analysis of the main scenarios and found some points of convergence (see Table 7.2).

[12] This analysis, conducted in 2013 for the year 2012, considered both the lower prices during daylight hours and the higher prices at night. The daily peak shaving allowed a saving estimated at 1.42 billion euros in 2012, more than three times the savings in 2011 (396 million). At the same time, however, there has been a rise in prices in the evening, when conventional power plants recover the income eroded by PV during the day. (QualEnergia 2013b).

Table 7.2. *Overview of key parameters under the IEA baseline scenario and under a scenario that meets the target of limiting CO_2 below 440 parts per million by 2050*

	Unit	2010	Baseline 2030	Baseline 2050	< 440 ppm 2030	< 440 ppm 2050
Global population	billion	7	8.3	9.1	8.3	9.1
Global GDP/ capita	thousand USD	10.9	17.4	24.3	17.4	24.3
Energy supply	EJ/year	530	645	749	474	407
Renewables	%	13	25	37	39	77
CO_2 emissions	Gt CO_2/year	27.4	38.5	56.6	36.7	7.1

Note: In the first scenario, the share of renewables in the energy mix will rise almost threefold, while in the second they will become the dominant energy source by 2050.
Sources: Fischedick et al. (2011) and IEA (2012a).

Currently, we harness less than 2.5 per cent of the available technical potential of renewable energy. All 150 available energy scenarios therefore agree that technology is not a serious constraint to the expansion of renewables. All of these scenarios foresee major growth of renewables, but no single renewable technology is expected to be dominant at a global level (Fischedick et al. 2011; IEA 2012a). Solar power has by far the largest potential, followed by geothermal, wind, and biomass (see Figure 7.9). Although the potential of bioenergy has been revised downward in recent years because of sustainability concerns (see Chapter 6), most scenarios indicate that bioenergy (mainly biofuels) will experience the most growth by 2050, followed by wind and solar.

Wind energy is currently the third most competitive renewable energy technology after hydropower and geothermal, and it has expanded rapidly over the last two decades in Europe and North America, and more recently in China and India. In order to achieve the high expansion indicated by most scenarios, greater geographical distribution will be needed. Yet this will be limited by environmental and NIMBY constraints. Offshore wind energy has a high potential, but further technological advances and major investment in grid connections between land and sea is needed before many of the most promising remote offshore locations can be exploited.

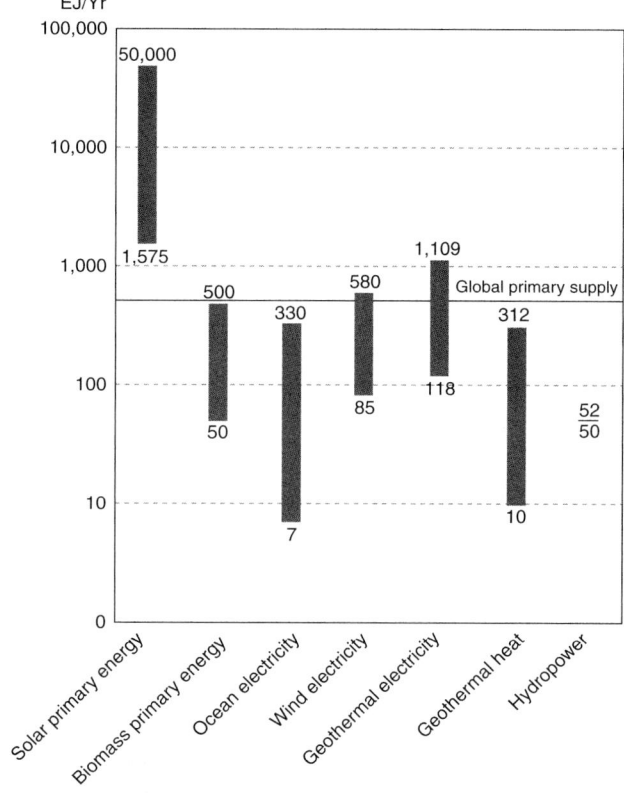

Figure 7.9. Ranges of global technical potentials of renewables compared with global energy supply and demands (logarithmic scale). Biomass and solar may have both heat and electricity uses and are considered primary energy sources.
Source: IPCC (2011) (modified).

Concentrating solar power (CSP), solar PV, geothermal heat pumps (GHPs), and enhanced geothermal systems (EGS) will require technological improvements and further improvements in cost efficiency before they achieve grid parity (are competitive) with fossil-based power.

Where will these alternative energies be deployed? Most scenarios suggest that developing countries will account for most of the increase in global energy demand over the coming decades, and that growth in renewable energy generation will also be highest in these countries. North Africa is the most interesting region for solar power, while North America has the greatest wind power potential.

256   The Renaissance of Renewable Energy

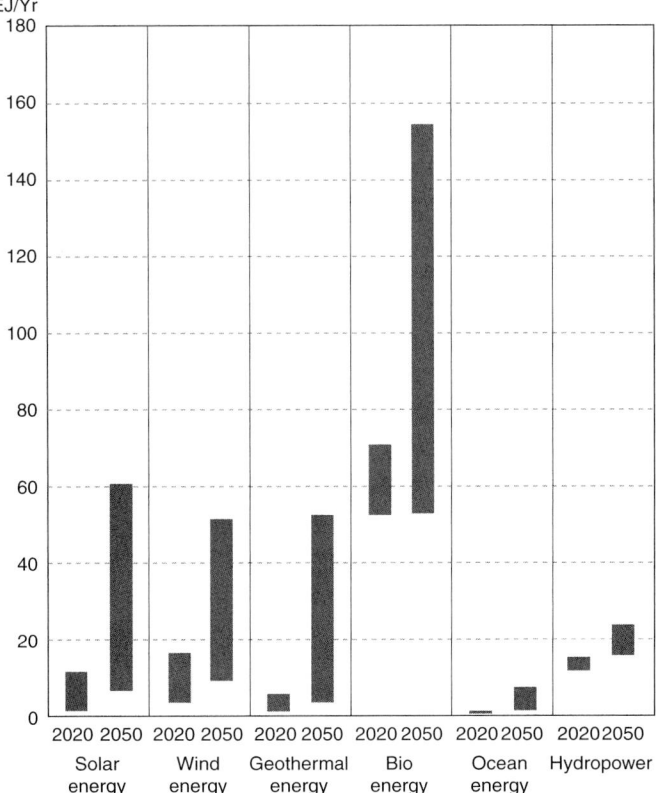

Figure 7.10. Projected ranges of global renewables use by source in 2020 and 2050. Total renewable energy deployment in 2050 is projected to be between 117 and 314 exajoules per year, while global primary production is expected to be between 600 and 1,500 exajoules per year. *Source:* Fischedick et al. (2011) (modified).

Large-scale hydropower potential has been largely exhausted in OECD countries, so most future expansion is expected to occur in Asia and Latin America.

7.7 Energy and Development

As a child growing up in Italy in the 1970s, I went to the cinema in my home city most Sundays. At that time the choice of snacks and soft drinks in the kiosk at the entrance was limited. Popcorn came in a 50-gram bag and soft drinks in a 200-milliliter glass bottle. Today, it's not unusual to see youngsters head into one of the ten theatres at the shiny new multiplex clutching a drum of popcorn large enough to bathe an infant in, and a litre of Coke in a giant plastic cup. Are the children with the supersize portions or greater choice of movies better off in any real sense? Might they actually be worse off? According to the World Health Organization, type 2 diabetes, largely the result of obesity and physical inactivity, claims 3.4 million lives per year and is likely to be the seventh leading cause of death by 2030. The two predicted leading causes of death by 2030 – heart disease and stroke – are also linked to poor eating and exercise habits (WHO 2013).

According to United Nations population projections, by 2050 the Earth will be home to more than 9 billion human beings. Nearly all the population increase during the next four decades will occur in developing countries, particularly Africa and South Asia. Urbanization will continue at an accelerated pace, reaching 70 per cent in 2050. Income levels will be multiples of what they are today, and in order to feed this larger, urban, and wealthier population, global food production will need to increase by 70 per cent (FAO 2009, 2011a).[13] This will require more cropland, water, fuels, buildings, industrial processes, transportation, and social services (Biello 2011b). It will also entail increased energy production, greater interference with the environment, and worrying social issues. More than a third of the world's people currently rely on wood, dung or animal waste to cook their food, and 20 per cent still lack access to electricity. Smoke inhalation from cooking with traditional biomass causes lung diseases that kill nearly 2 million people a year, most of them women and children. Avoidance of lung disease is not the only reason to move away from traditional biomass. Modern fuels for cooking and heating relieve women of the drudgery and dangers associated with foraging for wood, while electricity allows water to be pumped for crops, foods and medicines to be refrigerated, and children to study after dark.

The link between energy and development was explicitly defined at the World Summit on Sustainable Development in 2002: access to modern energy sources (such as electricity and natural gas) is crucial for the achievement of the UN's Millennium Development Goals.[14]

---

[13] To meet growing demand, annual cereal production will need to rise from the current yield of 2.1 billion tonnes to 3 billion tonnes, and annual meat production will need to increase from 200 million to 470 million tonnes (FAO 2009).

[14] The Millennium Development Goals are eight international development targets set by the United Nations in 2000 for achievement by 2015. These are: (1)

Figure 7.11. Barefoot College in Tilonia, a small village in Rajasthan (India), is a nongovernmental organization that has been providing basic services in rural communities for more than forty years, with the objective of making them self-sufficient and sustainable. In this class, thirty women learn how to install and maintain off-grid PV systems. *Source:* Gian Andrea Pagnoni.

Until the 1970s the assumption that economic development, measured by indicators such as gross domestic product (GDP),[15] ought to be the principal goal of all countries went largely unchallenged among economists. However, this runs counter to experience. The amount of money we spend raising children to adulthood indicates that most people value human relationships over money. At the societal level, it is also evident that improved economic conditions (such as during the 1950s in the United States) do not necessarily translate into higher levels of perceived happiness.

>   eradicating extreme poverty and hunger, (2) achieving universal primary education, (3) promoting gender equality and empowering women, (4) reducing child mortality rates, (5) improving maternal health, (6) combating HIV/AIDS, malaria and other diseases, (7) ensuring environmental sustainability and (8) developing a global partnership for development.
> [15] Gross domestic product is the market value of all goods and services produced within a country. Developed in the 1930s, GDP has become the most common way of measuring the size of a country's economy.

GDP is not a good indicator of quality of life for four main reasons. First, the GDP of a country may be high but its goods and services may be expensive, leaving many people badly off once they have paid for basic services such as health and education. Second, we tend to adapt quickly to new levels of prosperity. This is what happens to lottery winners, who, after an initial phase of elation, tend to return to their previous emotional state. Third, social dynamics come into play as perceived prosperity depends on comparison with others. Finally, although most people in industrialised countries have more leisure time than their grandparents did, their investment in relationships has, in many cases, decreased. Quality of life surveys show that participation in community life has declined in wealthier societies, and the number of people who die without the help of loved ones has increased (Becchetti 2012).

Recognising that numerous non-economic factors contribute to quality of life, the economists Ignacy Sachs and Amartya Sen helped to launch an alternative concept of development – human development – and an instrument to measure it, the Human Development Index (HDI). The HDI is currently based on three indicators: life expectancy at birth (a long and healthy life is representative of a higher level of development); expected years of schooling (education index); and

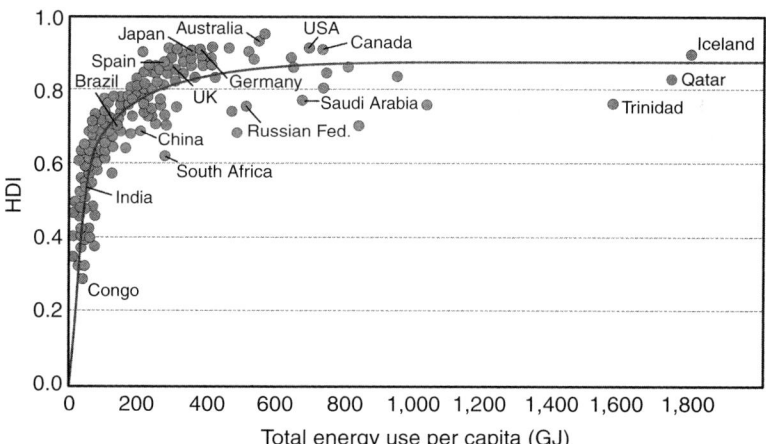

Figure 7.12. Relationship in 130 countries between energy use per capita (GJ per year) and HDI. This demonstrates that, above a certain level, energy consumption does not contribute to quality of life.
Sources: Sathaye et al. (2011) and Smil (2008), data from UNDP and World Bank.

gross national income per capita. Published annually by the United Nations, the HDI aims to support more rounded targets for development, beyond the merely economic.

If we look at the correlation between HDI and primary energy use per capita (Figure 7.12), it becomes clear that a minimum amount of energy (roughly 40 GJ per capita per year) is required to ensure a high standard of living (Sathaye et al. 2011). However, quality of life does not improve significantly above this level of consumption.

7.8 Energy Efficiency

While most people in the industrialised world agree in principle on the need to use energy more efficiently, relatively few understand how this can be achieved. A recent study by Columbia University in New York showed that 40 per cent of Americans believe that the most important way to save energy is to curtail energy-consuming activities, for example, driving less and turning lights off. Only 10 per cent identified what experts agree are the most effective measures in the long term: insulating homes and using more efficient appliances and vehicles (Attari et al. 2010).

This suggests that energy infrastructure matters more than consumer behaviour. One of the easiest places to start in improving energy efficiency is with buildings. The IEA estimates that we could improve the energy efficiency of buildings fivefold using existing techniques; in many cases, using techniques that have been around for centuries. For most of human civilization, the energy efficiency of housing has been far lower than it could have been, had the available techniques been applied (Smil 1994). There is no good reason why, for centuries, most people in cold and temperate zones lived in houses that were incapable of retaining heat. Growing up in Ireland, I experienced more than my share of draughts and winter evenings with a fire roasting my chest and an icy chill down my back. The cost of insulating an Irish house is far lower than the cost of the additional fuel required over the lifetime of that house. The failure to make insulation a mainstay of housing construction suggests a mixture of short-term planning and a preference, exhibited in much consumer behaviour, for the addition of a solution (in this case, fuel) over the removal of a problem (poor insulation).

This appears to be changing. Recent decades have already seen major improvements in building insulation standards. This has been most evident in central and northern Europe, most markedly with the development of the 'passive house' standard. A passive house is a building that requires no fuel-driven heating or cooling and has the lowest possible energy footprint for lighting, appliances and other energy needs. By combining passive solar heating techniques, such as triple glazing and heat recovery ventilation,[16] with active technologies, such as solar water heating, solar PV power generation and geothermal heating and cooling through heat pumps, a passive house completely eliminates its external energy requirements. Since 2007, the German city of Frankfurt has required that all new or restored public buildings meet passive house standards.

Energy efficiency has been catching on in other places, too. Three of the world's largest energy consumers – China, Japan and the United States – recently announced new energy efficiency measures.[17] However, even if these policies are implemented, they still leave enormous energy efficiency potential untapped (IEA 2012a).[18]

It is clearly important that we change our behaviour in relation to energy use, if only to prepare ourselves psychologically to accept larger structural changes to the energy economy (for example, recharging the car overnight rather than filling up with gas on demand). However, without structural changes, behaviour will have little impact on energy consumption. The more optimistic future energy scenarios assume that we will replace our existing infrastructure in the short term. For example, almost four-fifths of permissible carbon dioxide emissions (if we are to avoid global temperature increase above 2 degrees Celsius by 2035) are already locked in by existing power plants, factories, buildings and appliances. We must quickly replace the current infrastructures that waste energy in prodigious quantities at the points of conversion. There is little hope of staying below the 2-degree threshold unless we do so in the next five to ten years (IEA 2012a). This will require major investment by governments, private companies and, ultimately, citizens.

According to the IEA, "Energy efficiency is just as important as unconstrained energy supply, and increased action on efficiency can serve as a unifying energy policy that brings multiple benefits" (IEA

---

[16] Heat recovery ventilation transfers heat from the stale air expelled to the fresh air taken in.
[17] China is targeting a 16 per cent reduction in energy intensity by 2015; the United States has adopted new fuel economy standards; and Japan aims to cut its electricity consumption by 10 per cent by 2030 (IEA 2012a).
[18] Eighty per cent of the potential in the buildings sector and more than 50 per cent in industry (IEA 2012a).

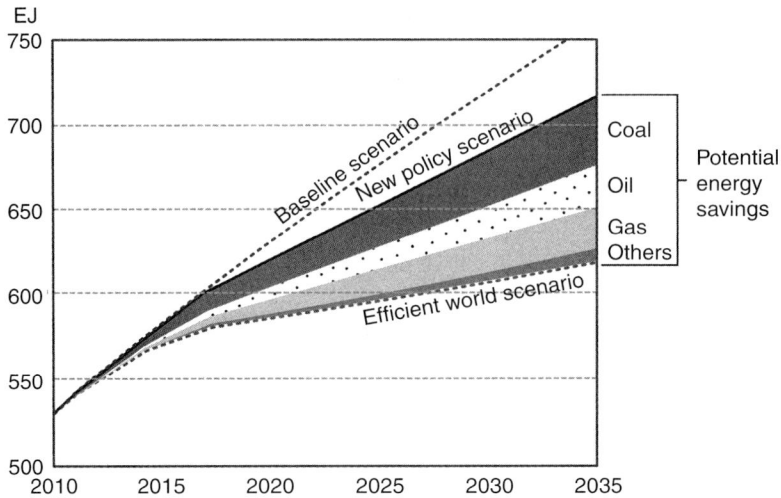

Figure 7.13. Total primary energy demand according to three different IEA scenarios. In addition to the business-as-usual (baseline) scenario, the IEA has provided a 'New Policy' scenario that takes account of policy commitments announced by governments but not yet implemented. The 'Efficient World' scenario shows how growth in energy demand up to 2035 can be limited through improvements in the efficiency of fossil-based technologies.
Source: IEA 2012a (modified).

2012b). Indeed by 2035, savings can amount to about 20 per cent of global energy consumption (see Figure 7.13) (IEA 2012a).

7.9 Switching from Oil to Electricity in Transportation

In the early pages of this book, we illustrated the problem of energy conversion with the example of a typical car, which converts only about 10 per cent of the energy contained in its fuel into motion. Although the internal combustion engine continues to be improved, its efficiency will always be severely hampered by its reliance on combustion and the attendant dispersion of energy. For that reason, any leap forward in terms of vehicle efficiency must involve a departure from the internal combustion engine and a re-embracing of the electric motor.

Had it not been for the discovery of cheap petroleum fuel, it is very likely that we would already have seen a steeper evolutionary curve for electric cars, since they have been around just as long as

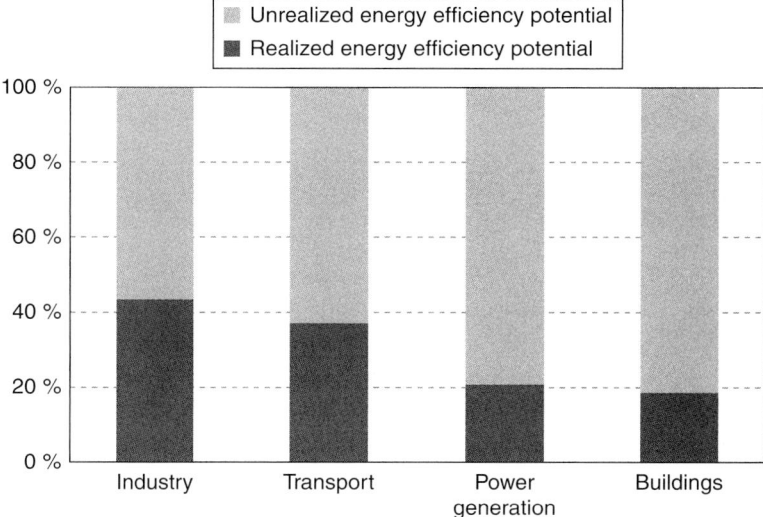

Figure 7.14. The potential for efficiency savings in each of the main energy sectors. According to the IEA, if the New Policies scenario is followed, by 2035 global industry will have met 40 per cent of its energy efficiency potential, while four-fifths of the potential to improve energy efficiency in the residential sector will remain untapped.
Source: IEA (2012a) (modified).

their gas-guzzling equivalents. As long as transport fuel remained relatively cheap, there was little incentive for companies, governments and, ultimately, consumers to invest in the research required to reach the holy grail of electric transportation: a small, light and powerful battery. That incentive has been growing steadily since the 1970s, augmented, in fact, by the sustainability imperative. Electric vehicles have not yet achieved the vehicular equivalent of grid parity, but many people believe that we are on the cusp of that moment.

Between 2011 and 2012, global sales of electric vehicles more than doubled, exceeding 180,000 units.[19] More than forty models of hybrid vehicle are available in the United States alone, and, prompted by policies to reduce GHG emissions, many more are likely to become available in the coming years. Most industrialised countries have introduced fuel economy targets, which put pressure on automakers

---

[19] Electric vehicles comprise plug-in hybrid electric vehicles, battery-electric vehicles, and fuel cell electric vehicles (Electric Vehicles Initiative 2013).

to produce more efficient vehicles. Some governments, such as that of California, have gone one step further by introducing quotas for zero-emissions vehicles (Pyper and ClimateWire 2012).[20]

Yet, electric vehicles still comprise only 0.02 per cent of all passenger cars (Electric Vehicles Initiative 2013), and two key challenges still remain to be overcome before that percentage is likely to increase significantly: infrastructure and battery power. Nowhere does the availability of charging stations come close to supporting the widespread adoption of electric vehicles. But even if there were electric charging stations on every street corner, without batteries that can be rapidly recharged there will be no way to compete with the convenience of a quick fill-up at a petrol station. If recharging times could be reduced to a matter of minutes rather than hours, this would allow motorists to combine a rest stop with a recharge, without any noticeable inconvenience vis-à-vis liquid fuels. Lithium-ion batteries currently represent the best hope, but they are also expensive. A recent survey conducted by consulting firm Pike Research found that consumers in the United States are willing to pay a premium for an electric vehicle; $23,750 for an electric vehicle comparable to a $20,000 gasoline-powered car. Unfortunately, all current electric models cost well above $30,000 (Pyper and ClimateWire 2012).

Of course, electric vehicles do not guarantee a clean and sustainable future, since they are only as clean or sustainable as the electricity they run on. According to Smil (2010), an all-electric fleet would not offer primary energy savings, and, unless it relies on renewable electricity, would not even offer carbon emission advantages.[21] Similarly, Demirdöven and Deutch, two researchers from the Massachusetts Institute of Technology, found that fuel cell vehicles using hydrogen from fossil fuels (see Chapter 2) offer no significant energy efficiency advantage over hybrid vehicles. They concluded that hybrid vehicles

---

[20] According to California's vehicle emissions targets, electric and fuel cell vehicles should make up about 15 per cent of its passenger vehicle fleet by 2025, and 90 per cent by 2050.

[21] There are 250 million passenger cars in the United States alone. If this entire fleet were to run on electricity, it would consume 25 per cent of the country's current power supply, generated largely through coal, natural gas and nuclear. If the efficiency of conventional power plants (40 per cent) and transmission losses from plant to battery are considered, every car would require 2 MJ of primary energy per kilometre of travel. This is equivalent to slightly more than 6 litres of gasoline per 100 kilometres, much less efficient than modern gasoline or hybrid motors (Smil 2010).

Figure 7.15. The filling station of the future? A PV sunshade that recharges electric vehicles.
Source: Tatmouss at Wikimedia Commons.

provided the most viable model for cleaner, more efficient transportation (Demirdöven and Deutch 2004).

7.10 The Sustainability 'Cure'

Sustainable development means different things in different places. In wealthy countries, it may mean cleaning up cities, encouraging more environmentally friendly forms of transport, or promoting renewable energy. In developing countries, it often involves choices between life and death. In its most literal interpretation, sustainable development is any economic development that can be sustained indefinitely, but its deeper meaning concerns the sustenance of the human race. The World Bank became a convert to the idea of sustainability in the early 1990s, claiming that "the achievement of a sustained and equitable development remains the greatest challenge facing the human race" (World Bank 1992, 1). It adopted a more pragmatic interpretation of sustainability than the UN Brundtland Report did, arguing that environmental protection need not be a priority, but should be factored into development projects, and that such projects may be deemed sustainable as long as they produce a net gain. Thus was born the idea of 'environmental accounting' – putting a price on the environment.

There may, ultimately, be a contradiction inherent in the juxtaposition of the terms 'sustainable' and 'development'. Sustainability is a dynamic concept that changes as the world changes. It is often understood as a synonym of green or renewable, yet it is, essentially, a philosophical concept. In this sense, sustainability might better be viewed as a process – a learning process – rather than a model for development.

Poverty is inextricably linked to sustainable development. Many of the world's poorest and most vulnerable people live in environmentally fragile regions. Unsustainable development therefore hits the poor first and hardest. In the climate change debate, lower-income countries sometimes claim they are being forced to choose between environmental protection and economic development. Yet, for development to be sustainable, it must actively reduce poverty. The problem of nonglobal solutions to the sustainability crisis is that many of the processes that produce pollution are being exported, often to the poorest countries.

The World Bank, UNEP, and UNDP set up the Global Environmental Facility in 1990 to provide funding for environmental projects such as emissions reduction, biodiversity protection, and water management. From 1991 to 2012 it distributed US$10.5 billion in grants and guaranteed US$51 billion in loans for more than 2,700 projects in more than 165 countries[22]. However, its impact, taken in the overall context of resource and capital flows between rich and poor countries, is small.

One of the most common rules of thumb in environmental protection is the 'precautionary principle', which is also enshrined in the Rio Declaration: "Where there are threats of serious or irreversible damage, lack of full scientific certainty shall not be used as a reason for postponing cost-effective measures to prevent environmental degradation" (United Nations 1992). The precautionary principle, like the term 'sustainable development' itself, is vague, leaving ample space for interpretation and political manoeuvring. How do we judge the seriousness of the threat if scientific evidence is lacking? Its critics argue that in reversing the burden of proof, such that one must show that an activity will not cause harm, the precautionary principle is inherently unscientific. Its proponents rely on the common sense

---

[22] See www.thegef.org

Figure 7.16. The sustainable housing project in Kronsberg, Germany. The heating requirements per household are 40 per cent below the German average, and the embodied carbon dioxide emissions in the buildings are almost 75 per cent lower than in conventional housing. *Source:* AxelHH at Wikimedia Commons.

argument: if you have strong reason to suspect that the box may be Pandora's, don't open it (at least not yet).

Cities present the greatest dilemma for sustainability practitioners. The only examples we have of truly sustainable human societies are low-density, land-based settlements, such as indigenous communities and modern eco-communities. Yet, with human population projected to stabilize at 9 billion around 2050, and with 6 billion of these people living in cities, it is clear that a return to the land is not on the cards (Blewitt 2008). Cities, as well as being a major part of the problem, will have to play a major role in the solution. Some attempts have been made to produce functioning models of a sustainable urban community. One of the most prominent of these is the suburb of Kronsberg in Hannover, Germany, which was developed specifically for the 2000 World Exposition as a blueprint for sustainable high-density development (see Figure 7.16). The core principles behind the settlement were energy, water, soil and soil management to minimize its environmental footprint. Today, the Expo Settlement in Kronsberg is a thriving community of 7,000 people.

Another model that has gathered momentum is the transition town movement. Pioneered in the English town of Totnes, it mobilizes urban dwellers in existing settlements to move away from energy-intensive lifestyles. Rob Hopkins, one of the movement's founders, sees urban sustainability in terms of energy: it is about coping with "a continual decline in the net energy supporting humanity" (Blewitt 2008, 162). He sees the solution in communities becoming more localised and self-reliant. The transition town movement was inspired by the example of Cuba, where, following the end of the Cold War and the loss of Russian oil supplies, agriculture was de-intensified and localised. Hopkins sees "peak energy as an opportunity for positive change rather than as an inevitable disaster" and describes a path of 'energy descent', where communities adapt by developing energy alternatives and reducing consumption (Blewitt 2008, 162). This is the essence of the transition movement – managing a peaceful and crisis-free transition to a post–fossil fuel age.

Indigenous cultures might be described as the global specialists in sustainability. Many of the core principles of ecology – interdependence, ecological cycles, energy flows, diversity, complexity and co-evolution – are already deeply rooted in indigenous cultures. Australia's aboriginal people managed to live sensitively with the rhythm and dynamics of the Earth for 40,000 years – leaving a barely discernible footprint and dispensing with the need for leaders and hierarchies. In their culture, shared responsibility trumped the exercise of power, both over humans and the environment (Blewitt 2008).

Some indigenous peoples, such as the Ashanínka in the Amazon, have begun to share their expertise in sustainable land management with outsiders. Their lands and survival threatened by loggers, ranchers and mining companies, one Ashanínka community living along the Amônia River in western Brazil founded a training centre called Yorenka Ãtame (meaning 'knowledge of the rainforest') in 2007 (see Figure 7.17). Its aim is to teach non-indigenous people about sustainable forest management techniques, such as those successfully practiced by the Ashanínka for generations. This project involves a transfer of knowledge from indigenous to non-indigenous people, reversing the direction of information flow that has been the norm for centuries.

Sustainable management of our planet's resources requires that we recognise ecosystems as the basic units that we have to be able to live within. In order for the ecosystem approach to succeed, new types of collaboration and partnership will be necessary – between civil society, the private sector, and governments (WWF 2008).

Figure 7.17. The Brazilian minister of culture, Juca Ferreira, visits the Yorenka Ãtame training centre with representatives of the Ashanínka people and local government officials.
Source: Pedro França/Ministério da Cultura do Brasil.

7.11 The New Energy Experience: From NIMBY to YIMBY

Renewable energy is technically capable of meeting the energy needs of the human race. The limits are set not by technology, but by economic, environmental, political, and social constraints. Of these, the economic and social constraints are the most potent. Since most renewable technologies are not yet cost competitive with conventional energy sources, they will need subsidies to encourage their deployment until they reach grid parity. These are already in place in several countries. Moreover, technological improvement and economies of scale will continue to reduce unit costs. Finally, if measures to internalise the pollution costs of conventional fuels – such as a carbon tax – are implemented as planned, the playing field between conventional and alternative energy will level and eventually tip in favour of cleaner energy.

The social barriers to renewables are less tangible but represent perhaps the most important key to unlocking this vast natural resource. Though most people feel vaguely uneasy about the existing energy system, it has one enormous advantage over the alternatives: familiarity. Though proverb tells us that familiarity breeds contempt, psychology holds the opposite to be true. Numerous studies have shown that the more exposure we have to a stimulus, the more we will tend to like it. The word itself holds a clue. What is more familiar to us than our family? We may sometimes feel that we hate them as much as we love them, but in the end we usually stick by them.

The power available at the flick of a switch from a faceless utility, the filling station where we can also, conveniently, pick up a cold drink, the power stations comfortably hidden from view – these things are part of people's expectation of the world, at least in those countries fortunate enough to enjoy plentiful access to energy. The prospect of switching to a vastly different system is frightening. The NIMBY attitude to energy infrastructure shows that many people don't want to think too deeply about where their energy comes from. Yet, the inexorable reality of global warming impresses upon us that the entire world is our back yard. It doesn't matter whether carbon dioxide is emitted a mile from our home or ten thousand miles away; we will feel its impact either way.

One of the key differences between conventional and renewable energies is that the latter do not lend themselves to massive hidden facilities. On the contrary, the great advantage of renewable energy is that we *can* install it in our backyards. It invites us into a relationship with our energy source. This may explain some of the resistance to renewable energy (for example, to highly visible facilities such as wind farms), but is also one of its greatest selling points. As we have shown in earlier chapters, the detachment we feel from our energy source is very recent. Our ancestors used to live with or next to the animals that tilled their fields, and they felled the timber that heated their homes. There is no way we can return to this idyll, if it can even be described as such. The world's population is expected to stabilizes at 9 billion or 10 billion people over the next century. Most of these people will live in cities or suburbs.

There are ways to engage more closely with your energy source, without actually installing a biogas digester or wind turbine in your yard, or placing solar panels on your roof. Throughout Europe, more than 2,400 renewable energy cooperatives have been set up, with funding from the European Union, allowing citizens to invest and acquire a financial stake in the energy they use. According to Adrien Bullier of the European Commission, "We see great potential in this kind of model. It fosters acceptability" (Hall 2014).

The solution to our future energy needs will probably be renewable (otherwise the Earth will be too hot to inhabit), and almost certainly far more decentralised than the current system. That means, both urban and rural communities will have to become accustomed to the sight of energy infrastructure – solar water heaters, PV panels, wind turbines, biogas plants, geothermal plants, small-scale hydro, and a variety of new delivery mechanisms – literally and figuratively in their backyards.

The renaissance of renewable energy will therefore require from citizens both a more localised and a more global understanding of energy. Whether it is generated on our roof, ten miles offshore, or on another continent, the energy we use will have to be our business. We will cease to be mere consumers and start to hold a stake in where our energy comes from. We will need to develop a positive attitude to energy, as something we have chosen and rightfully own. It's about saying, "Yes, in my back yard."

References

Adams, W. M., 2004. *Against Extinction: The Story of Conservation*. London: Earthscan.
Al-Riffai, P., B. Dimaranan, and D. Laborde, 2010. *Global Trade and Environmental Impact Study of the EU Biofuels Mandate*. Report of International Food Policy Research Institute, March. http://www.ifpri.org/sites/default/files/publications/biofuelsreportec.pdf.
Al Seadi, T., and C. Lukehurst, 2012. "Quality Management of Digestate from Biogas Plants Used as Fertiliser." *IEA Bioenergy Task* 37. www.iea-biogas.net.
Alsema, E. A., M. J. de Wild-Scholten, and V. M. Fthenakis, 2006. "Environmental Impacts of PV Electricity Generation. A Critical Comparison of Energy Supply Options." Presented at the 21st European Photovoltaic Solar Energy Conference, Dresden, Germany, September 4–8. http://www.clca.columbia.edu/papers/21-EUPVSC-Alsema-DeWild-Fthenakis.pdf.
Andrews, C. J., L. Dewey-Mattia, J. M. Schechtman, and M. Mayr, 2011. "Alternative Energy Sources and Land Use." In *Climate Change and Land Policies*, edited by Gregory K. Ingram and Yu-Hung Hong, 89–115. Cambridge: Lincoln Institute of Land Policy.
Archer, C. L., and M. Z. Jacobson, 2005. "Evaluation of Global Wind Power." *Journal of Geophysical Research*. 110:D12110.
Arvizu, D., P. Balaya, L. Cabeza, T. Hollands, A. Jäger-Waldau, M. Kondo, et al. 2011a. "Direct Solar Energy." In IPCC 2011.
Arvizu, D., T. Bruckner, H. Chum, O. Edenhofer, S. Estefen, et al., 2011b. "Technical Summary." In IPCC 2011.
Attari, S. Z., M. L. DeKay, and W. B. de Bruin, 2010. "Public Perceptions of Energy Consumption and Savings." www.pnas.org/cgi/doi/10.1073/pnas.1001509107.
Ausubel, J., 2007. "Renewable and Nuclear Heresies." *International Journal of Nuclear Governance, Economy and Ecology* 1 (3).
Baffes, J., and T. Haniotis, 2010. *Placing the 2006/08 Commodity Price Boom into Perspective*. World Bank Development Prospects Group. Policy Research Working Paper 5371. July.
Bain, R. L., 2011. *Biopower Technologies in Renewable Electricity Alternative Futures*. Golden, CO: National Renewable Energy Laboratory.
Becchetti, L., 2012. "La felicità interna lorda: la crisi oltre il PIL." Lecture of a professor of economics at the University of Rome Tor Vergata. http://www.economia.rai.it/articoli/la-felicit%C3%A0-interna-lorda-la-crisi-oltre-il-pil/18700/default.aspx.

Behr, P., and ClimateWire, 2011. "Smart Grid Works for Utilities, but Not Yet for Consumers." *Scientific American*, April 13. http://www.scientificamerican.com/article.cfm?id=smart-grid-works-utilities-not-yet-consumers.

Benner, J. H. B., Berkhuizen, J. C., de Graaff, R. J., and Postma, A. D., 1993. Impact of the wind turbines on birdlife. Final report no. 9247. Consultants on Energy and the Environment. Rotterdam, The Netherlands.

Bertani, R., 2012. "Geothermal Power Generation in the World 2005–2010 Update Report." *Geothermics* 41:1–29.

Biello, D., 2009. "The Origin of Oxygen in Earth's Atmosphere." *Scientific American*. August 19. http://www.scientificamerican.com/article.cfm?id=origin-of-oxygen-in-atmosphere.

Biello, D., 2010. "The Price of Coal in China: Can China Fuel Growth without Warming the World? China Relies on Coal for Most of Its Energy while Striving to Cut Greenhouse Gas Emissions and Other Pollution." *Scientific American*. December 16. http://www.scientificamerican.com/article.cfm?id=price-of-coal-in-china-climate-change.

Biello, D., 2011a. "The False Promise of Biofuels. The Breakthroughs Needed to Replace Oil with Plant-Based Fuels Are Proving Difficult to Achieve." *Scientific American*, August 10. http://www.scientificamerican.com/article.cfm?id=the-false-promise-of-biofuels.

Biello, D., 2011b. "Human Population Reaches 7 Billion – How Did This Happen and Can It Go On?" *Scientific American*. October 28. http://www.scientificamerican.com/article.cfm?id=human-population-reaches-seven-billion.

Biello, D., 2012a. "Craig Venter Explains How Pond Scum Will Save the World." *Scientific American*, January 5. http://www.scientificamerican.com/article.cfm?id=pond-scum-to-the-rescue&WT.mc_id=SA_CAT_ENGYSUS_20120112.

Biello, D., 2012b. "Pay Dirt: How to Turn Tar Sands into Oil. More and More Petroleum Is Flowing from Alberta's Vast Oil Sands Deposits." *Scientific American*, December 10. http://www.scientificamerican.com/article.cfm?id=how-to-turn-tar-sands-into-oil-slideshow&WT.mc_id=SA_CAT_ENGYSUS_20121213.

Biello, D., 2013a. "EPA on Keystone XL: Significant Climate Impacts from Tar Sands Pipeline." *Scientific American*, April 23. http://blogs.scientificamerican.com/observations/2013/04/23/epa-on-keystone-xl-pipeline-environmental-impacts/.

Biello, D., 2013b. "Fracking Could Help Geothermal Become a Power Player." *Scientific American*, July 29. http://www.scientificamerican.com/article.cfm?id=fracking-for-renewable-power-geothermal&WT.mc_id=SA_CAT_ENGYSUS_20130801.

Biello, D., 2013c. "Oil Sands Raise Levels of Cancer-Causing Compounds in Regional Waters." *Scientific American*, January 8. http://www.scientificamerican.com/article.cfm?id=oil-sands-raise-levels-of-carcinogens-in-regional-waters&WT.mc_id=SA_CAT_ENGYSUS_20130110.

Biello, D., 2013d. "The Opposite of Mining: Tar Sands Steam Extraction Lessens Footprint, but Environmental Costs Remain." *Scientific American*, January 2. http://www.scientificamerican.com/article.cfm?id=tar-sands-extraction-without-strip-mining&WT.mc_id=SA_CAT_ENGYSUS_20130103.

Biello, D., 2013e. "Will Alternative Energy Growth Tank during New Fossil-Fuel Glut?" *Scientific American*, March 13. http://www.scientificamerican.com/article.cfm?id=alternative-energy-challenged-by-abundant-fossil-fuels.

Biello, D., 2014. "Fight over Rooftop Solar Forecasts a Bright Future for Cleaner Energy." *Scientific American*, March 25. http://www.scientificamerican.com/article/fight-over-rooftop-solar-forecasts-a-bright-future-for-cleaner-energy/?&WT.mc_id=SA_ENGYSUS_20140327.

Blewitt, J., 2008. *Understanding Sustainable Development*. London: Earthscan.
Boehlert, G. W., and A. B. Gill, 2010. "Environmental and Ecological Effects of Ocean Renewable Energy Development, a Current Synthesis." *Oceanography* 23 (2): 68–81.
BP, 2011. "BP Statistical Review of World Energy." bp.com/statisticalreview.
BP, 2012. "BP Statistical Review of World Energy." bp.com/statisticalreview.
Braun, R., P. Wiland, and A. Wellinger, 2009. "Biogas from Energy Crop Digestion. IEA Bioenergy. Task 37 – Energy from Biogas and Landfill Gas." www.iea-biogas.net.
Brittaine, R., and N. Lutaladio, 2010. "Jatropha: A Smallholder Bioenergy Crop. The Potential for Pro-Poor Development." *Integrated Crop Management* 8. http://www.fao.org/docrep/012/i1219e/i1219e.pdf.
Brown, T. A., M. K. Jones, W. Powell, and R. G. Allaby, 2008. "The Complex Origins of Domesticated Crops in the Fertile Crescent." *Trends in Ecology and Evolution* 24 (2): 103–109.
Bruckner, T., H. Chum, A. Jäger-Waldau, Å. Killingtveit, L. Gutiérrez-Negrín, J. Nyboer, et al. (eds)], 2011. Cambridge University Press, Cambridge, United Kingdom and New York, NY, USA.
Brumfiel, G., 2012. "Nuclear Fusion Project Struggles to Put the Pieces Together." *Scientific American and Nature*, October 26. http://www.scientificamerican.com/article.cfm?id=nuclear-fusion-project-struggles-to-put-the-pieces-together.
Brumfiel, G., 2013. "Fukushima: Fallout of Fear. After the Fukushima Nuclear Disaster, Japan Kept People Safe from the Physical Effects of Radiation – but Not from the Psychological Impacts." *Nature*, January 16. http://www.nature.com/news/fukushima-fallout-of-fear-1.12194.
Bullis, K., 2012. "Sapphire Energy Raises $144 Million for an Algae Farm" *MIT Technology Review*, April 6. http://www.technologyreview.com/news/427431/sapphire-energy-raises-144-million-for-an-algae-farm/.
Carson, R., 2002. *Silent Spring*. New York: Mariner Books. First Published 1962
Cataldi, R., 1999. "The Year Zero of Geothermics." In *Stories from a Heated Earth*, edited by R. Cataldi, S. Hodgson, and J. W. Lund, 7–17. Sacramento, CA: Geothermal Resources Council and International Geothermal Association.
Cavallin, F., A. Lorenzoni, and G. Sofia, 2011. "Prospettive di parità" *Qualenergia*, Novembre-Dicembre 2011, 36–40. www.qualenergia.it.
Chen, Y., A. Ebenstein, M. Greenstone, and H. Li, 2013. "Evidence on the Impact of Sustained Exposure to Air Pollution on Life Expectancy from China's Huai River Policy." http://www.pnas.org/content/early/2013/07/03/1300018110.
Chum, H., A. Faaij, J. Moreira, G. Berndes, P. Dhamija et al. 2011. "Bioenergy." In IPCC 2011.
Coleridge, S. T., 1857. *The Rime of the Ancient Mariner*. New York: D. Appleton & Co.
Collins, P. G., 2005. "Making Cold Antimatter." *Scientific American* (June): 78–85.
Connor, S., 2009. "Nuclear power? Yes please … Exclusive: leading greens join forces in a major U-turn." The Independent, 23 February. http://www.independent.co.uk/environment/green-living/nuclear-power-yes-please-1629327.html
Cooper, M., 2009. "The Economics of Nuclear Reactors: Renaissance or Relapse?" Paper by Senior Fellow for Economic Analysis Institute for Energy and the Environment Vermont Law School. www.vermontlaw.edu/documents/cooper%20report%20on%20nuclear%20economics%20final%5B1%5D.pdf.

Crawford, R. H., 2009. "Lifecycle Energy and Greenhouse Emissions Analysis of Wind Turbines and the Effect of Size on Energy Yield." *Renewable and Sustainable Energy Reviews* 13 (9): 2653–2660.

Dale, M., S. Krumdieck, and P. Bodger, 2011. "Net Energy Yield from Production of Conventional Oil." *Energy Policy* 39:7095–7102.

da Rosa, A., 2005. *Fundamentals of Renewable Energy Processes*. Amsterdam: Elsevier.

David, M. J., S. Bowman, J. K. Balch, P. A. William, J. Bond et al., 2009. "Fire in the Earth System." *Science* 324 (5926): 481–484.

de Castro, C., M. Mediavilla, L. J. Miguel, and F. Frechoso, 2011. "Global Wind Power Potential: Physical and Technological Limits." *Energy Policy* 39:6677–6682.

de Lucas, M., Janss, G. F. E., and Ferrer, M. (Ed.), 2007. *Birds and Wind Farms: Risk Assessment and Mitigation*. Quercus: Madrid.

Demirbas, A., 2010. "Use of Algae as Biofuel Sources." *Energy Conversion and Management* 51 (12): 2738–2749.

Demirdöven, N., and J. Deutch, 2004. "Hybrid Cars Now, Fuel Cell Cars Later." *Science* 305:974–976.

DESERTEC, 2009. *Clean Power from Deserts, the DESERTEC Concept for Energy, Water and Climate Security*. 4th ed. Bonn: Protext Verlag http://www.desertec.org/en/global-mission/milestones/.

Desideri, U., S. Proietti, F. Zepparelli, P. Sdringola, and S. Bini, 2012. "Life Cycle Assessment of a Ground-Mounted 1778 kWp Photovoltaic Plant and Comparison with Traditional Energy Production Systems." *Applied Energy* 97: 930–943.

Deutch, J. M., and E. J. Moniz, 2006. "The Nuclear Option." *Scientific American* 295 (3): 76–83.

Dickens, C. 1995. *Hard Times – For These Times*. London: Penguin Books. First Published 1854. p. 27

Dickson, M. H. and M. Fanelli, 2004. "What is Geothermal Energy?" http://www.geothermal-energy.org/geothermal_energy/what_is_geothermal_energy.html#c344

Didier, F., 2011. "Solar Photovoltaics Competing in the Energy Sector: Grid Parity and Beyond." http://www.epia.org/?id=968.

Doom, J., and E. Goossens, 2012. "China Says U.S. Renewable Subsidies Violate Trade Rules." *Bloomberg*, May 24. http://www.bloomberg.com/news/2012-05-24/china-says-u-s-renewable-subsidies-violate-trade-rules.html.

Ehrlich, P. R., 1968. *The Population Bomb*. New York: Sierra Club/Ballantine Books.

Eisentraut, A., 2010. *Sustainable Production of Second-Generation Biofuels Potential and Perspectives in Major Economies and Developing Countries*. IEA Information Paper. www.iea.org.

Electric Vehicles Initiative, 2013. "Global EV Outlook: Understanding the Electric Vehicle Landscape to 2020". http://www.iea.org/publications/globalevoutlook_2013.pdf

Energia24, 2012. "Il boom del fotovoltaico ha trasformato il sistema elettrico nazionale. 10/05/2012." http://energia24club.it/01NET/HP/0,1254,51_ART_147625,00.html.

Erickson, W., G. Johnson, and D. Young, 2005. "A Summary of Bird Mortality from Anthropogenic Causes with an Emphasis on Collisions." USDA Forest Services Gen. Tech. PSW-GRT-191.

ETSAP, 2010. "Marine Energy Technology Brief E13 – November, 2010". Energy Technology Systems Analysis Programme, International Energy Agency, Paris, France. www.etsap.org/E-techDS/PDF/E08-Ocean%20Energy_GSgct_Ana_LCPL_rev30Nov2010.pdf.

EU press release, 2012. "New Commission Proposal to Minimise the Climate Impacts of Biofuel Production. Brussels, 17 October 2012." http://europa.eu/rapid/press-release_IP-12-1112_en.htm.

Eurobserver, 2013. "Photovoltaic Barometer." http://www.energies-renouvelables.org/observ-er/stat_baro/observ/baro-jdp9.pdf.

Everaert, J., and Kuijken, E. 2007. *Wind Turbines and Birds in Flanders (Belgium)*. Brussels: Research Institute for Nature and Forest (INBO) http://www.windaction.org/documents/11725

Fabbri, C., and M. Soldano, 2010. "Biometano, un'alternativa al biogas in cogenerazione." *Agricoltura, settembre* 2010:92–94.

Fairley, P., 2011. "Next Generation Biofuels." *Nature* 474:2–5.

FAO, 2001. Human energy requirements. Report of a Joint FAO/WHO/UNU Expert Consultation. Rome, 17–24 October 2001.

FAO, 2009. "How to Feed the World in 2050. High-Level Experts Forum, Rome on 12–13 October 2009." http://www.fao.org/fileadmin/templates/wsfs/docs/expert_paper/How_to_Feed_the_World_in_2050.pdf.

FAO, 2011a. "Agriculture Key to Addressing Future Water and Energy Needs." http://www.fao.org/news/story/en/item/94760/icode/.

FAO, 2011b. "'Energy-Smart' Food for People and Climate." http://www.fao.org/docrep/014/i2454e/i2454e00.pdf.

FAO/WHO/UNU, 2001. *Human Energy Requirements*. Report of a Joint FAO/WHO/UNU Expert Consultation. Rome, October 17–24. ftp://ftp.fao.org/docrep/fao/007/y5686e/y5686e00.pdf.

Fargione, J. H., J. Tilman, D. Polasky, and S. Hawthorne, 2008. "Land Clearing and the Biofuel Carbon Debt." *Science* 319:1235–1238.

Feynman, R. P., 1970. *The Feynman Lectures on Physics: The Definitive and Extended Edition*. 3 vols., 2nd ed. Boston: Addison Wesley.

Fischedick, M., R. Schaeffer, A. Adedoyin, M. Akai, T. Bruckner et al., 2011. *Mitigation Potential and Costs. In IPCC Special Report on Renewable Energy Sources and Climate Change Mitigation*. Cambridge: Cambridge University Press.

Fisk, R., 2006. *The Great War for Civilisation: The Conquest of the Middle East*. New York: HarperCollins.

Fortune Magazine, 2012. "Global 500. Annual Ranking of the World's Largest Corporations." http://money.cnn.com/magazines/fortune/global500/2012/full_list/.

Gibbs, W. W., 2009. "Plan B for Energy: 8 Revolutionary Energy Sources." *Scientific American*, April 2. http://www.scientificamerican.com/article.cfm?id=plan-b-for-energy-8-ideas&page=2.

Goldstein, B., G. Hiriart, R. Bertani, C. Bromley, L. Gutiérrez-Negrín, E. Huenges, H. Muraoka, A. Ragnarsson, J. Tester, and V. Zui, 2011. "Geothermal Energy." In IPCC 2011.

Gowrisankaran, G., S. S. Reynolds, and M. Samano, 2011. "Intermittency and the Value of Renewable Energy." http://sfb-seminar.uni-mannheim.de/material/renewable_intermittency.pdf.

Graham-Rowe, D., 2011. "Beyond Food versus Fuel." *Nature* 474:6–8.

Greenpeace, 2012. "Fukushima Nuclear Disaster: Who Profits and Who Pays?" http://www.greenpeace.org/international/en/news/Blogs/makingwaves/fukushima-who-profits-who-pays/blog/40463/.

Greenpeace, 2013. "Fukushima Fallout. Nuclear Business Makes People Pay and Suffer." http://www.greenpeace.org/africa/Global/international/publications/nuclear/2013/FukushimaFallout.pdf.

Groat, C. G., and T. W. Grimshaw, 2012. *Fact-Based Regulation for Environmental Protection in Shale Gas Development*. Austin: Energy Institute. University of Texas at Austin. http://energy.utexas.edu/images/ei_shale_gas_regulation120215.pdf.

Hacatoglu, K., M. A. Rosen, and I. Dincer, 2012. "Comparative Life Cycle Assessment of Hydrogen and Other Selected Fuels." *International Journal of Hydrogen Energy* 37:9933–9940.

Hadhazy, A., 2009. "Will Space-Based Solar Power Finally See the Light of Day?" *Scientific American*, April 16. http://www.scientificamerican.com/article.cfm?id=will-space-based-solar-power-finally-see-the-light-of-day&page=4.

Hall, C. A. S., and K. A. Klitgaard, 2012. *Energy and the Wealth of Nations: Understanding the Biophysical Economy*. Dordrecht: Springer.

Hall, M., 2014. "Europeans test energy cooperatives to boost independence." Euractiv.com, 30 June. http://www.euractiv.com/sections/energy/europeans-test-energy-cooperatives-boost-independence-303145

Hansen, J., 2008. "4th Generation Nuclear Power." http://ossfoundation.us/projects/energy/nuclear.

Harkki, S., 2012. "Food, Fuel, Forests and Climate – the Biofuels Conundrum. October 18, 2012." http://www.greenpeace.org/international/en/news/Blogs/makingwaves/food-fuel-forests-and-climate-the-biofuels-co/blog/42642/.

Haszeldine, R. S., 2009. "Carbon Capture and Storage: How Green Can Black Be?" *Science* 325:1647–1652.

Higgins, P., 2007. "The Origins of Hydroelectricity." *Ecologist*, September 6. http://www.theecologist.org/investigations/energy/269238/the_origins_of_hydroelectricity.html.

Hodson Peter, V., 2012. "History of Environmental Contamination by Oil Sands Extraction." www.pnas.org/cgi/doi/10.1073/pnas.1221660110.

Holland-Bartels, L., and B. Pierce, 2011. "An Evaluation of the Science Needs to Inform Decisions on Outer Continental Shelf Energy Development in the Chukchi and Beaufort Seas, Alaska. U.S. Department of the Interior and U.S. Geological Survey." http://pubs.usgs.gov/circ/1370/pdf/circ1370.pdf.

Howard, B., L. Parshall, J. Thompson, S. Hammer, J. Dickinson, and V. Modi, 2012. "Spatial Distribution of Urban Building Energy Consumption by End Use." *Energy and Buildings* 45:141–151. Amsterdam: Elsevier

Hubbard, B., 2012. "Wind Turbines: The Future of Renewable Energy or a Blight on UK Countryside?" *Ecologist*, February 24. http://www.theecologist.org/investigations/energy/1258482/wind_turbines_the_future_of_renewable_energy_or_a_blight_on_uk_countryside.html.

Huber, G. W., and B. E. Dale, 2009. "Grassoline at the Pump." *Scientific American*, July 1.

Hume, D., 1737. *An Enquiry Concerning Human Understanding*. 2012 Edition. Hamburg: Tredition.

Hvistendahl, M., 2008. "China's Three Gorges Dam: An Environmental Catastrophe?" *Scientific American*, March 25. http://www.scientificamerican.com/article.cfm?id=chinas-three-gorges-dam-disaster.

IAEA, 2010. "International Status and Prospects of Nuclear Power. 2010 Edition." http://www.iaea.org/newscen.

IAEA, 2012. "Nuclear Technology Review." http://www.iaea.org/Publications/Reports/ntr2012.pdf.

IEA, 2005. *Energy Statistics Manual*. OECD/IEA, Paris, France.

IEA, 2007. "Bioenergy Project Development. Biomass Supply. Good Practice Guidelines." www.iea.org.

IEA, 2008. "Deploying Renewables. Principles for Effective Policies." www.iea.org.

IEA, 2009a. "Renewable Energy Essentials: Concentrating Solar Thermal Power." http://www.iea.org/publications/freepublications/publication/CSP_Essentials.pdf.

IEA, 2009b. "Technology Roadmaps Carbon Capture and Storage." http://www.iea.org/publications/freepublications/publication/name,3847,en.html.

IEA, 2009c. "Wind Energy Roadmap Targets." http://www.iea.org/publications/freepublications/publication/Wind_Roadmap_targets_viewing.pdf.

IEA, 2010a. "Projected Costs of Generating Electricity 2010 Edition." www.iea.org/textbase/nppdf/free/2010/projected_costs.pdf.

IEA, 2010b. "Technology Roadmap. Concentrating Solar Power." http://www.iea.org/papers/2010/csp_roadmap.pdf.

IEA, 2012a. *World Energy Outlook 2012*. http://www.iea.org/publications/freepublications/publication/WEO2012_free.pdf

IEA, 2012b. "North America leads shift in global energy balance, IEA says in latest World Energy Outlook." Press Release, 12 November. http://www.iea.org/newsroomandevents/pressreleases/2012/november/name,33015,en.html

IEA Bioenergy, 2009. "Bioenergy, a Sustainable and Reliable Energy Source. A Review of Status and Prospects. Main Report." http://www.ieabioenergy.com/LibItem.aspx?id=6479.

IEA Geothermal, 2012. "Trends in Geothermal Applications. Survey Report on Geothermal Utilization and Development in IEA-GIA Member Countries in 2010 with Trends in Geothermal Power Generation and Heat Use 2000–2010. Publication of the IEA Geothermal Implementing Agreement, July 2012." www.iea-gia.org

IJHD, 2010. *World Atlas & Industry Guide. International Journal of Hydropower and Dams*. Wallington, Surrey: IJHD.

Inman, M., 2013a. "How to Measure the True Cost of Fossil Fuels." *Scientific American*, April 1.

Inman, M., 2013b. "Will Fossil Fuels Be Able to Maintain Economic Growth? A Q&A with Charles Hall." *Scientific American*, March 21. http://www.scientificamerican.com/article.cfm?id=eroi-charles-hall-will-fossil-fuels-maintain-economic-growth.

IPCC, 2007. "IPCC Fourth Assessment Report (AR4) 'Climate Change 2007.'" http://www.ipcc.ch/publications_and_data/publications_and_data_reports.shtml#.UnvlueKDpQU.

IPCC, 2011. *IPCC Special Report on Renewable Energy Sources and Climate Change Mitigation*. Edited by O. Edenhofer, R. Pichs-Madruga, Y. Sokona, K. Seyboth, P. Matschoss, S. Kadner, T. Zwickel, P. Eickemeier, G. Hansen, S. Schlömer, and C. von Stechow. Cambridge: Cambridge University Press.

IPCC, 2013. "Climate Change 2013: The Physical Science Basis." http://www.ipcc.ch/report/ar5/wg1/#.UnvOaeKDpQX.

Irfan, U., and ClimateWire, 2013. "Return of the Hydrogen Car? An Idled Program for Vehicles Driven by Fuel Cells Is Gaining New Impetus from New Programs at the U.S. Department of Energy." *Scientific American*, June 4. http://www.scientificamerican.com/article.cfm?id=return-of-the-hydrogen-car&WT.mc_id=SA_CAT_ENGYSUS_20130606.

Isler, K., and C. P. van Schaik, 2009. "The Expensive Brain: A Framework for Explaining Evolutionary Changes in Brain Size." *Journal of Human Evolution* 57:392–400.

Jacobson, M. Z. and M. A. Delucchi, 2009. A Plan to Power 100 Percent of the Planet with Renewables. Scientific American November 2009.

Jacobson, M. Z., and M. A. Delucchi, 2011a. "Providing All Global Energy with Wind, Water, and Solar Power, Part I: Technologies, Energy Resources, Quantities and Areas of Infrastructure, and Materials." *Energy Policy* 39:1154–1169.

Jacobson, M. Z., and M. A. Delucchi, 2011b. "Providing All Global Energy with Wind, Water, and Solar Power, Part II: Reliability, System and Transmission Costs, and Policies." *Energy Policy* 39:1170–1190.

Janss, G., A. Lazo, J. M. Baqués, and M. Ferrer, 2001. Some Evidence of Changes in Use of Space by Raptors as a Result of the Construction of a Wind Farm, 2001. Paper presented at 4th Eurasian Congress on Raptors. September 25–29, Seville, Spain.

JRC, 2011. "Technology Map of the European Strategic Energy Technology Plan." http://setis.ec.europa.eu/about-setis/technology-map/2011_Technology_Map1.pdf.

Kanter, J., 2012. "Obstacles to Danish Wind Power." *New York Times*, January 22. http://www.nytimes.com/2012/01/23/business/global/obstacles-to-danish-wind-power.html?_r=0.

Kavar, T., and P. Dovč, 2008. "Domestication of the Horse: Genetic Relationships between Domestic and Wild Horses." *Livestock Science* 116:1–14.

Ketterer, D. (Ed.), 1984. *Mark Twain. Tales of Wonder*. Lincoln: University of Nebraska Press.

King, C. W., and C. A. Hall, 2011. *"Relating Financial and Energy Return on Investment." Sustainability* 3 (10): 1810–1832.

Kingsbury, K., 2007. "After the Oil Crisis, a Food Crisis?" *Time*, November 16. http://www.time.com/time/business/article/0,8599,1684910,00.html?iid=sphere-inline-sidebar.

Kojima, R., 1995. "Urbanization in China." *Developing Economies* 33:151–154.

Kopsakangas-Savolainen, M., and R. Svento, 2012. *Modern Energy Markets*. Dordrecht: Springer.

Krewitt, W., K. Nienhaus, C. Kleßmann, C. Capone, E. Stricker, W. Graus, M. Hoogwijk, N. Supersberger, U. von Winterfeld, and S. Samadi, 2009. *Role and Potential of Renewable Energy and Energy Efficiency for Global Energy Supply*. Climate Change 18/2009, ISSN 1862–4359. Dessau-Roßlau, Germany: Federal Environment Agency.

Kubiszewski, I., C. J. Cleveland, and P. K. Endres, 2010. "Meta-Analysis of Net Energy Return for Wind Power Systems." *Renewable Energy* 35:218–225.

Kumar, A., T. Schei, A. Ahenkorah, R. Caceres Rodriguez, J. M. Devernay, M. Freitas, D. Hall, Å. Killingtveit, and Z. Liu, 2011. "Hydropower." In IPCC 2011.

Kümmel, R., 2011. *The Second Law of Economics. Energy, Entropy, and the Origins of Wealth*. Dordrecht: Springer.

Kump, L. R., 2011. "L'ultimo grande riscaldamento globale." *Le Scienze Settembre* 2011:51–55.

Kurek, J., L. K. J. Derek, C. G. Muir, X. Wang, M. S. Evans, and J. P. Smol, 2012. "Legacy of a Half Century of Athabasca Oil Sands Development Recorded by Lake Ecosystems. Proc Natl Acad Sci USA, 10.1073/pnas.1217675110." http://intl.pnas.org/content/early/2013/01/02/1217675110.abstract.

Lake, J. A., R. G. Bennett, and J. F. Kotek, 2009."Next Generation Nuclear Power." *Scientific American*, January 26. http://www.scientificamerican.com/article.cfm?id=next-generation-nuclear.

Lal, B. and P. M. Sarma, 2011. *Wealth from Waste*. Third Edition. New Delhi: TERI Press.

Larsson, A., 1989. "State of the Art Report on Radioactive Waste Disposal. Different Types of Radioactive Wastes Can Be, and Are Being, Stored and Disposed of Safely." *IAEA Bulletin* 4:18–25. http://www.iaea.org/Publications/Magazines/Bulletin/Bull314/31402691825.pdf.

Leitenberg, M., 2006. *Deaths in Wars and Conflicts in the 20th Century*. Cornell University Peace Studies Program. Occasional paper 29/3rd edition. http://www.isn.ethz.ch/Digital-Library/Publications/Detail/?id=16951

Lewis, A., S. Estefen, J. Huckerby, W. Musial, T. Pontes, and J. Torres-Martinez, 2011. "Ocean Energy." In IPCC 2011.

Liu, C., 2011. "Climate Change Evaporates Part of China's Hydropower. The Nation's Hydropower Production Dropped by 25 Percent Thanks to an Unusual Drop in River Flow." http://www.scientificamerican.com/article.cfm?id=climate-change-evaporates-china-hydropower-production-drop-25-percent&WT.mc_id=SA_CAT_ENGYSUS_20111110.

Lovejoy, C. O., 1988. "Evolution of Human Walking." *Scientific American* 259 (5): 82–89.

Lovelock, J., 2007. *The Revenge of Gaia: Why the Earth Is Fighting Back – and How We Can Still Save Humanity*. London: Penguin Books.

Lund, J. W., 2007. Characteristics, development and utilization of geothermal resources. Geo Heat Center Bulletin, June 2007. http://geoheat.oit.edu/pdf/tp126.pdf

Marks J. C., 2007. Down go the dams. *Scientific American*, March 2007:65–71.

Marquart, M., 2011. "Energiewende: Russland bietet Deutschen milliardenschweren Pakt an." *Der Spiegel*, November 14. http://www.spiegel.de/wirtschaft/soziales/energiewende-russland-bietet-deutschen-milliardenschweren-pakt-an-a-797549.html

Masden, E. A., A. D. Fox, R. W. Furness, R. Bullman, and D. T. Haydon, 2010. "Cumulative impact assessments and bird/wind farm interactions: Developing a conceptual framework." *Environmental Impact Assessment Review* 30: 1–7.

Massoud, A., and P. F. Schewe, 2008. "Preventing Blackouts: Building a Smarter Power Grid." *Scientific American*, August 13. http://www.scientificamerican.com/article.cfm?id=preventing-blackouts-power-grid.

Maugeri, L., 2009. "Squeezing More Oil Out of the Ground." *Scientific American*, April 1. http://www.scientificamerican.com/article/squeezing-more-oil-edit-this/

McDonald, R. I., J. Fargione, J. Kiesecker, W. M. Miller, and J. Powell, 2009. "Energy Sprawl or Energy Efficiency: Climate Policy Impacts on Natural Habitat for the United States of America." *PLoS One* 4 (8): 1–11. http://www.plosone.org/article/info:doi/10.1371/journal.pone.0006802.

McGowin, C., 2008. Renewable Energy Technical Assessment Guide. TAG-RE: 2007, Electric Power Research Institute, Palo Alto, CA, USA.

McHenry, H. M, 2009. "Human Evolution." In *Evolution: The First Four Billion Years*, edited by Michael Ruse and Joseph Travis, 263. Cambridge, MA: Belknap Press of Harvard University Press.

McNeill, J. R., 2000. *Something New under the Sun: An Environmental History of the Twentieth-Century World*. New York: W. W. Norton.

MEA (Millennium Ecosystem Assessment), 2005. Ecosystems and Human Well-being. Washington, DC: Synthesis. Island Press. http://www.millenniumassessment.org/documents/document.356.aspx.pdf

Meadows, D., D. Meadows, J. Randers, and W. Behrens III, 1974. *The Limits to Growth: A Report for the Club of Rome's Project on the Predicament of Mankind*. New York: Signet Books. First published 1972.

Melillo, J. M., J. M. Reilly, D. W. Kicklighter, A. C. Gurgel, T. W. Cronin, et al., 2009. "Indirect Emissions from Biofuels: How Important?" *Science* 326:1397–1399.

Miller, L. M., F. Gans, and A. Kleidon, 2011. "Estimating Maximum Global Land Surface Wind Power Extractability and Associated Climatic Consequences." *Earth Systems Dynamics* 2: 1–12.

Meneguzzo, F., 2011. "Conoscere per decidere: elettricità fotovoltaica, elettricità nucleare e mercato elettrico." http://www.aspoitalia.it/attachments/301_F_Meneguzzo%20-costo%20kWh%20FV%20-7%20giugno%202011-1.pdf.

Moerschbaecher, M., and J. W. Day Jr., 2011. "Ultra-Deepwater Gulf of Mexico Oil and Gas: Energy Return on Financial Investment and a Preliminary Assessment of Energy Return on Energy Investment." *Sustainability* 3:2009–2026. http://www.mdpi.com/2071–1050/3/10/2009.

Moomaw, W., F. Yamba, M. Kamimoto, L. Maurice, J. Nyboer, et al., 2011. *Introduction*. In IPCC 2011.

Morales, A., and G. Sulugiuc, 2011. "Mossi & Ghisolfi Starts to Build Bio-Ethanol Plant." *Bloomberg*, April 13. http://www.bloomberg.com/news/2011-04-12/construction-begins-on-first-commercial-cellulosic-ethanol-plant-in-italy.html.

Morris, C., 2013. "Negative Power Prices on the Weekend. Renewables International 17/06/2013." http://www.renewablesinternational.net/negative-power-prices-on-the-weekend/150/537/67152/.

Moyer, M., 2010. "Fusion's False Dawn." *Scientific American*, March 1.

Murphy, D. J., and C. A. S. Hall, 2010. "Year in Review – EROI or Energy Return on (Energy) Invested." *Annals of the New York Academy of Sciences* 1185:102–118.

Murphy, J., R. Braun, P. Weiland, and A. Wellinger, 2011. "Biogas from Crop Digestion." *Energy IEA Bioenergy Task* 37. www.iea-biogas.net.

Murray, J., and D. King, 2012. "Oil's Tipping Point Has Passed. 26 January 2012, National Energy Policy, 2001. Reliable, Affordable, and Environmentally Sound Energy for America's Future. Report of the National Energy Policy Development Group. May 2001." *Nature* 481:433–435.

NAE, 2000. "Greatest Engineering Achievements of the 20th Century." http://www.greatachievements.org.

Nawaz, I., and G. N. Tiwari, 2006. "Embodied Energy Analysis of Photovoltaic Based on Macro- and Micro-Level." *Energy Policy* 34:3144–3152.

Niele, F., 2005. *Energy: Engine of Evolution*. Amsterdam: Elsevier.

NREL, 2012. *Renewable Energy Data Book*. National Renewable Energy Laboratory (US Dept. of Energy), Golden, CO, USA. http://www.nrel.gov/docs/fy14osti/60197.pdf

OECD/IEA, 2008. "Turning a Liability into an Asset: Landfill Methane Utilisation Potential in India." http://www.iea.org/publications/freepublications/publication/name,3785,en.html.

OECD/IEA, 2009. "Turning a Liability into an Asset: the Importance of Policy in Fostering Landfill Gas Use Worldwide." http://www.iea.org/publications/freepublications/publication/name,3800,en.html.

OECD/IAEA, 2010. "Uranium 2009. Resources, Production and Demand." http://www.oecdbookshop.org/oecd/display.asp?lang=EN&sf1=identifiers&st1=978-92-64-04789-1.

Ombello, C., 2010. "The World's First Molten Salt Concentrating Solar Power Plant." *Guardian*, July 22. http://www.guardian.co.uk/environment/2010/jul/22/first-molten-salt-solar-power.

Palmer, D., 2010. "U.S. Challenges China Wind Power Aid at WTO." *Reuters*, December 22. http://www.reuters.com/article/2010/12/22/us-usa-china-windpower-idUSTRE6BL3EU20101222.

Percival, S. M., 2007. "Predicting the effects of wind farms on birds in the UK: the development of an objective assessment method." In: de Lucas, M. et al. (Ed.) *Birds and Wind Farms: Risk Assessment and Mitigation.* pp. 137–152.

Petersson, A., and A. Wellinger, 2009. "Biogas Upgrading Technologies, Developments and Innovations. Task 37 – Energy from Biogas and Landfill Gas." www.iea-biogas.net.

Pfund, N., and B. Healey, 2011. "What Would Jefferson Do? The Historical Role of Federal Subsidies in Shaping America's Energy Future. DBL Investors." http://i.bnet.com/blogs/dbl_energy_subsidies_paper.pdf.

Platt, J. R., 2009. "Gorillas versus Charcoal: Biomass to the Rescue." *Scientific American*, April 23. http://blogs.scientificamerican.com/extinction-countdown/2009/04/23/gorillas-versus-charcoal-biomass-to-the-rescue/.

Pollard, K. S., 2009. "What Makes Us Human?" *Scientific American*, April 20. http://www.scientificamerican.com/article/what-makes-us-human/

Pottinger, P., 2012. "Community Energy: A Powerful Force." *International Rivers*, August 23. http://www.internationalrivers.org/resources/community-energy-a-powerful-force-7645.

Pyper, J., and ClimateWire, 2012. "Electric Vehicles Proliferate, but Prove a Tough Sell." *Scientific American*, September 24. http://www.scientificamerican.com/article.cfm?id=electric-vehicles-proliferate-but-prove-a-tough-sell.

QualEnergia, 2013a. "Domenica 16 giugno, rinnovabili al 100% e il prezzo dell'elettricità va a zero." http://qualenergia.it/articoli/20130617-domenica-16-giugno-prezzo-di-acquisto-elettricit%C3%A0-pari-a-zero.

QualEnergia, 2013b. "Quei 50 miliardi che le rinnovabili farebbero guadagnare all'Italia." http://qualenergia.it/articoli/20130418-quei-50-miliardi-che-le-rinnovabili-farebbero-guadagnare-a-italia.

Raugei, M., P. Fullana-i-Palmer, and V. Fthenakis, 2012. "The Energy Return on Energy Investment (EROI) of Photovoltaics: Methodology and Comparisons with Fossil Fuel Life Cycles." *Energy Policy* 45:576–582.

REN21, 2011. "Renewables 2011. Global Status Report." www.ren21.net

REN21, 2012. "Renewables 2012. Global Status Report." www.ren21.net.

Rhodes, R., 2010. *Twilight of the Bombs: Recent Challenges, New Dangers, and the Prospects for a World Without Nuclear Weapons.* New York: Alfred A. Knopf.

Richter, C., S. Teske, and Rebecca Short, 2009. Concentrating Solar Power. Global Outlook 2009 "Why Renewable Energy Is Hot.". Amsterdam/Tabernas/Brussels: Greenpeace International, SolarPACES, and ESTELA. http://www.greenpeace.org/international/en/publications/reports/concentrating-solar-power-2009/.

Rigden, J. S., 2005. *Einstein 1905: The Standard of Greatness.* Cambridge: Harvard University Press.

Roberts, P., 2005. *The End of Oil; on the Edge of a Perilous New World.* Boston: Mariner Books.

Rochon, E., E. Bjureby, P. Johnston, R. Oakley, D. Santillo et al., 2008. *False Hope. Why Carbon Capture and Storage Won't Save the Climate.* Amsterdam: Greenpeace International. http://www.greenpeace.org/international/en/publications/reports/false-hope/.

Rodrigue, J. P., C. Comtois and B. Slack, 2013. *The Geography of Transport Systems.* New York: Routledge.

Rubbia, C., 2012. "Energies for a Sustainable Future." Presentation to BSI Gamma Foundation, June 1. Potsdam: Institute for Advanced Sustainability Studies. https://www.bsibank.com/dms/site-bsi/docs/Gamma-Foundation-PDFs/2012/Energies-for-a-sustainable-Future---Carlo-Rubbia/Energies%20for%20a%20sustainable%20Future%20-%20Carlo%20Rubbia.pdf.

Rybach, L., 2005. "The advance of geothermal heat pumps world-wide." IEA Heat Pump Centre Newsletter 23, 13–18. http://www.geoexchangebc.com/pdfs/news_IEA_De05c.pdf

Saeta, P. N., 1999. "What Is the Current Scientific Thinking on Cold Fusion? Is There Any Possible Validity to This Phenomenon?" *Scientific American*, October 21. http://www.scientificamerican.com/article.cfm?id=what-is-the-current-scien.

Sanderson, K., 2011. "A Chewy Problem. The Inedible Parts of Plants Are Feeding the Next Generation of Biofuels. But Extracting the Energy-Containing Molecules is a Challenging Task." *Nature* 474, no.7352, p. S12–S14.

Sari, A. P., M. Maulidya, R. N. Butarbutar, R. E. Sari, and W. Rusmantoro, 2007. *Indonesia and Climate Change*. Working Paper on Current Status and Policies. Report of Department for International Development (DFID) and World Bank. http://siteresources.worldbank.org/INTINDONESIA/Resources/226271-1170911056314/3428109-1174614780539/PEACEClimateChange.pdf.

Sathaye, J., O. Lucon, A. Rahman, J. Christensen, F. Denton, J. Fujino, G. Heath, S. Kadner, M. Mirza, H. Rudnick, A. Schlaepfer, and A. Shmakin, 2011. "Renewable Energy in the Context of Sustainable Development." In IPCC 2011.

Savage, N., 2011. "The Scum Solution. The Green Slime That Covers Ponds Is an Efficient Factory for Turning Sunlight into Fuel, but Growing It on an Industrial Scale Will Take Ingenuity." *Nature* 474:S15–S16.

Scientific American, 2012. "Stop Burning Rain Forests for Palm Oil." *Scientific American*, December 6. http://www.scientificamerican.com/article.cfm?id=stop-burning-rain-forests-for-palm-oil&WT.mc_id=SA_CAT_ENGYSUS_20121206.

Shahan, Z., 2012. "China to Have 3 GW of Concentrated Solar Thermal Power (CSP) by 2020." *Scientific American*, March. http://www.scientificamerican.com/article.cfm?id=china-to-have-3-gw-of-concentrated-2012-03.

Sierra Club, 2012. "Background: Environmental Impacts of Tar Sands Development." http://www.sierraclub.org/energy/factsheets/tarsands.asp.

Singh, J., and S. Gu, 2010. "Commercialization Potential of Microalgae for Biofuels Production." *Renewable and Sustainable Energy Reviews* 14:2596–2610.

Sinha, P., C. J. Kriegner, W. A. Schew, S. W. Kaczmar, M. Traister, and D. J. Wilson, 2008. "Regulatory Policy Governing Cadmium-Telluride Photovoltaics: A Case Study Contrasting Life-Cycle Management with the Precautionary Principle." *Energy Policy* 36:381–387.

Smil, V., 1994. *Energy in World History*. Boulder: Westview Press.

Smil, V., 2000. *Energies: An Illustrated Guide to the Biosphere and Civilization*. Cambridge: The MIT Press.

Smil, V., 2006. *Energy. A Beginner's Guide*. London: Oneworld.

Smil, V., 2008. *Energy in Nature and Society. General Energetics of Complex Systems*. Cambridge: The MIT Press.

Smil, V., 2010. *Energy Myths and Realities. Bringing Science to the Energy Policy Debate*. Washington: AEI Press.

Smil, V., 2012. "A Skeptic Looks at Alternative Energy. It Takes Several Lifetimes to Put a New Energy System into Place, and Wishful Thinking Can't Speed Things Along." *ieee spectrum*, July. http://spectrum.ieee.org/energy/renewables/a-skeptic-looks-at-alternative-energy/0#.

Smith, A. H. V., 1997. "Provenance of Coals from Roman Sites in England and Wales." *Britannia* 28:297–324.

Soccol Carlos, R., L. P. Vandenberghe, A. B. Medeiros, S. G. Karp, M. Buckeridge, et al., 2010. "Bioethanol from lignocelluloses: Status and Perspectives in Brazil." *Bioresource Technology* 101(13): 4820–4825.

Sorensen, B., 2004. *Renewable Energy*. Third Edition. Amsterdam: Elsevier.

Sterner, D., Orloff, S., and Spiegel, L., 2007. "Wind turbine collision research in the United States." In: de Lucas, M. et al. (Ed.), *Birds and Wind Farms: Risk Assessment and Mitigation*. pp. 81–100.

Stix, G., 2006. "A Climate Repair Manual. Global Warming Is a Reality. Innovation in Energy Technology and Policy Are Sorely Needed If We Are to Cope." *Scientific American* 295 (3): 46–49.

Strickland, M. D., E. B. Arnett, W. P. Erickson, D. H. Johnson, G. D. Johnson et al., 2011. *Comprehensive Guide to Studying Wind Energy/Wildlife Interactions. Prepared for the National Wind Coordinating Collaborative*. Prepared for the National Wind Coordinating Collaborative, Washington, D.C., USA.

Swann, G. M. P., 2009. *The Economics of Innovation*. Cheltenham: Edward Elgar Publishing.

Timoney, K. P, and P. Lee, 2009. "Does the Alberta Tar Sands Industry Pollute? The Scientific Evidence." *Open Conservation Biology Journal* 3:65–81.

Tollefson, J., N. Gilbert, and *Nature* magazine, 2012. "Earth Summit: A Report Card to Preview the Rio+20 Mega-Conference." *Scientific American*, June 18. http://www.scientificamerican.com/article.cfm?id=report-card-preview-rio-20-mega-conference.

Tvedt, T., 2013. *A Journey in the Future of Water*. New York: I.B. Tauris.

United Nations, 1987. Report of the World Commission on Environment and Development: Our Common Future. New York: United Nations. http://www.un-documents.net/our-common-future.pdf

United Nations, 1992. Report of the United Nations Conference on Environment and Development (Rio de Janeiro, 3–14 June 1992). New York: United Nations. http://www.un.org/documents/ga/conf151/aconf15126-1annex1.htm

UN, 2007. "Sustainable Consumption and Production. Promoting Climate-Friendly Household Consumption Patterns. Prepared by the United Nations Department of Economic and Social Affairs." New York: United Nations. http://www.un.org/esa/sustdev/publications/household_consumption.pdf.

UNDP, 2012. "Rural Energy Development Programme in Nepal." http://www.undp.org/content/undp/en/home/ourwork/environmentandenergy/projects_and_initiatives/rural-energy-nepal/.

UNFPA, 2011. "The State of World Population 2011." http://foweb.unfpa.org/SWP2011/reports/EN-SWOP2011-FINAL.pdf

U.S. Congress Budget Office, 2003. "Energy Policy Act of 2003. Congressional Budget Office. S.14." www.cbo.gov/sites/default/files/cbofiles/ftpdocs/42xx/doc4206/s14.pdf.

Vallee, B. L., 1998. "Alcohol in the Western World." *Scientific American*, June 1.

Verein der Kohlenimporteure e.V., 2010. "Annual Report. Facts and Trends 2009/2010." http://www.verein-kohlenimporteure.de/wEnglish/download/vdki_2010_Engl_internet.pdf?navid=15.

Victor, D. G., and J. C. House, 2006. "BP's Emissions Trading System." *Energy Policy* 34:2100–2112.

Wald, M. L., 2009. "Is There a Place for Nuclear Waste?" *Scientific American*, August.

Weiss, W., and F. Mauthner, 2011. "Solar Heat Worldwide. Markets and Contribution to the Energy Supply 2009. AEE – Institute for Sustainable Technologies and IEA Solar Heating & Cooling Programme, May 2011." http://www.iea-shc.org/publications/downloads/Solar_Heat_Worldwide-2011.pdf.

White, L. A., 1943. "Energy and the Evolution of Culture." *American Anthropologist New Series* 45 (3): 335–356.

WHO, 2011. "Tackling the Global Clean Air Challenge." News release. 26 September, Geneva. http://www.who.int/mediacentre/news/releases/2011/air_pollution_20110926/en/index.html.

WHO, 2013. "Diabetes. Fact Sheet N° 312." http://www.who.int/mediacentre/factsheets/fs312/en/index.html.

WHO/UNDP, 2009. "The Energy Access Situation in Developing Countries. A Review Focusing on the Least Developed Countries and Sub-Saharan Africa." http://www.who.int/indoorair/publications/energyaccesssituation/en/index.html.

Williams, E., 2004. "Energy Intensity of Computer Manufacturing: Hybrid Assessment Combining Process and Economic Input–Output Methods." *Environmental Science & Technology* 38 (22): 6166–6174.

Wiser, R., Z. Yang, M. Hand, O. Hohmeyer, D. Infield, P. H. Jensen, V. Nikolaev, M. O'Malley, G. Sinden, and A. Zervos, 2011. "Wind Energy." In IPCC 2011.

WNA, 2005. *The New Economics of Nuclear Power*. World Nuclear Association Report. http://www.world-nuclear.org/reference/pdf/economics.pdf.

Wong, K., 2006. "Lucy's Baby. An Extraordinary New Human Fossil Renews Debate over the Evolution of Upright Walking." *Scientific American*, December 1.

World Bank, 1992. *World Development Report 1992: Development and the Environment*. Oxford: Oxford University Press.

World Bank, 2008. *Agriculture for Development. World Development Report 2008*. Washington, DC: World Bank. http://siteresources.worldbank.org/INTWDR2008/Resources/2795087-1192111580172/WDROver2008-ENG.pdf.

WWF, 2008. *Living Planet Report 2008*. Gland: WWF International.

Yee, A., 2012. "Microhydro Drives Change in Rural Nepal." *New York Times*, June 20. http://www.nytimes.com/2012/06/21/business/global/microhydro-drives-change-in-rural-nepal.html?_r=1&src=recg.

Yergin, D., 2011. *The Quest: Energy, Security and the Making of the Modern World*. New York: Penguin Press.

Zamparelli, C., 2005a. "Analisi critica degli specchi ustori di Archimede." http://www.mondosolare.it/pub/specchi2.pdf.

Zamparelli, C., 2005b. "Storia, scienza e leggenda degli specchi ustori di Archimede." http://www.gses.it/pub/specchi1.pdf.

Zecca, A., and C. Zulberti, 2006. "Petrolio: siamo in riserva?" *Le Scienze*, November 2006 (Italian edition of Scientific American).

Zoback, M., S. Kitasei, and B. Copithorne, 2010. *Addressing the Environmental Risks from Shale Gas Development*. Worldwatch Institute Briefing Paper. http://www.worldwatch.org/files/pdf/Hydraulic%20Fracturing%20Paper.pdf.

Index

aerodynamics, 85, 86
agriculture, 15, 17, 18, 19, 56, 57, 67, 69, 128, 140, 145, 146, 148, 181, 185, 189, 195, 212, 214, 239, 268
   history of, 190
   wastes, 139
Amundsen, Roald, 177
anaerobic digestion, 132, 134, 135, 136, 140
animals
   use of animal power, 17, 18-19, 22, 56, 190
antimatter, 169
Archimedes, 119, 122
Aristotle, 3, 6
Armenia, 140
Armstrong, William, 76, 235
artificial fertilizers, 56, 190
Aswan Dam, 205, 206
Atlantic Empress oil spill, 198
Atomic Energy Commission (United States), 25
Australia, 118, 124, 212, 227, 241, 249, 268
Austria, 102, 136
Ausubel, Jesse, 185
autotrophs, 12, 13, 14
Azerbaijan, 197

Bailey, William J., 95
Bangladesh, 83
Banqiao Dam disaster, 205
base load (electricity), 72, 73, 125, 156, 168, 186
batteries, 10, 11, 39, 40, 42, 63, 104, 105, 116, 162, 263, 264
Beaufort scale, 90
Becquerel, Edmond, 104
Belgium, 91, 119, 186
Benedick, Richard, 184

Bethe, Hans, 166
Bin Talal, Prince Hassan, 230
biofuels, 26, 28, 38, 62, 67, 116, 126-148, 215, 216, 221, 239
   algae-based, 143
   biodiesel, 38, 62, 127, 128, 139, 140, 141, 213, 215
   bioethanol, 38, 127, 128, 140, 215, 250
   biogas, 33, 73, 127, 131, 132, 135, 136, 139, 140, 145, 271
   controversy surrounding, 139
   economics of, 144-145
   environmental impacts of, 213
   ethanol, 38, 62, 138, 139, 141, 142, 146, 193
   first-generation, 139-140
   potential of, 145-148
   second-generation, 140-142
   sugarcane, 138
   third-generation, 142-143
biogas
   bacterial, 128
biomass, 12, 14, 17, 19, 20, 25, 28, 59, 67, 69, 70, 73, 75, 103, 126, 128, 129, 131, 132, 135, 138, 140, 144, 145, 148, 185, 213, 218, 220, 221, 233, 235, 250, 254, 257
bitumen, 35, 36, 200, 228, 235
Boormann, Frank, 183
BP (British Petroleum), 52, 63, 64, 65, 187, 211, 227, 236
Brazil, 38, 64, 82, 102, 135, 138, 139, 144, 145, 146, 176, 184, 189, 215, 221, 268
Brundtland Report, 28, 183, 265
Brundtland, Gro Harlem, 183

Canada, 36, 64, 102, 161, 177, 198, 226, 235, 237, 251

287

Cancún Agreements, 241, 265
capacity factor, 72, 248
cap-and-trade, 226
carbon, 10, 12, 13, 18, 29, 34, 106,
    135, 194, 233
  footprint, 181
  pricing, 145, 185, 269
  trading, 226–227
carbon capture and storage (CCS),
    219, 220, 229
carbon dioxide, 10, 12, 34, 38, 106,
    128, 143, 179, 180, 184, 194,
    203, 216, 219, 220, 227, 229,
    240, 241
  absorption of, 213
  emissions of, 117, 219, 236, 239,
    261, 264, 270
  in fermentation, 13
  in photosynthesis, 12
Caribbean, 85
cars, 5, 7, 8, 12, 23, 32, 47, 60, 61, 63,
    138, 139, 140, 219, 230, 232,
    262, 264
  electric cars, 12, 39, 40, 92, 244,
    263, 264
  embodied energy of, 61
  flexible fuel vehicles (FFVs), 139
  fuel cell vehicles, 264
  hybrid vehicles, 63, 263, 264
Carson, Rachel, 182
Carter, Jimmy, 245
Caucasus, 22, 174
cellulose, 126, 131, 132, 141
CERN (European Organization for
    Nuclear Research), 170
Charanka Solar Park, 116
charcoal, 18, 127, 129, 130, 145, 213
chemical reaction, 10, 106
Chernobyl, 187
Chile, 212
China, 17, 39, 47, 60, 64, 69, 77, 82,
    84, 93, 101, 116, 118, 124, 135,
    138, 145, 173, 176, 177, 181,
    184, 189, 194, 205, 212, 240,
    250, 254, 261
China (ancient), 23, 149
chlorofluorocarbons (CFCs), 184
Churchill, Winston, 174
Clean Air Act (US), 182
Clean Development Mechanism
    (United Nations), 135, 227
climate change, 25, 140, 176–181,
    184, 185, 200, 203, 216, 223,
    231, 233, 236, 240, 241, 266
  scenarios, 241
coal, 18, 20, 21, 23, 26, 27, 31, 32, 33,
    34, 35, 51, 52, 62, 70, 71, 75,
    82, 120, 137, 175, 176, 177,
    181, 191, 193, 194, 195, 196,
    201, 216, 221, 233, 235, 237,
    245, 249
  and railways, 21
  anthracite, 34, 64
  brown coal (lignite), 64
  economics of, 186
  historical use of, 21
  Industrial Revolution, 19
  reserves, 64
  resurgence of, 173
  use in shipping, 174
coke, 27, 189
Cold War, 185, 268
combined heat and power (CHP), 127
combustion
  cellular, 12
  chemical reaction, 27
computers
  embodied energy of, 61
Congo, 213
Conti, Prince Piero Ginori, 150
Cuba, 268
Czech Republic, 119

Da Vinci, Leonardo, 119
Darrieus wind turbine, 87
Davis, George, 205
De Rivaz, François Isaac, 42
Deepwater Horizon oil spill, 187, 198
deforestation, 20, 179, 185, 213
Delphi, Oracle of, 33
Delucchi, Mark, 242
Denmark, 92, 251
Descartes, René, 119
DESERTEC Foundation, 125, 242
DESERTEC project, 124, 125, 230,
    242, 244
Dickens, Charles, 189
diesel, 8, 11, 12, 32, 38, 42, 62, 96,
    138, 140
Diesel, Rudolf, 138
district heating, 128, 144, 192
dynamo, 70, 89

earthquakes, 11, 76, 188
ecological footprint, 240
Edison, Thomas, 24, 76
Egypt, 69, 249
Ehrlich, Paul, 179
Einstein, Albert, 18, 25, 29, 106, 165
electric motor, 5, 8, 63, 262
electricity, 8, 10, 11, 12, 22, 23, 25,
    26, 34, 42, 52, 59, 69, 70, 72,
    73, 80, 82, 85, 89, 90, 92, 93,
    105, 117, 121, 122, 123, 124,
    125, 144, 160, 250
  access to, 257
  advantages of, 24
  advent of, 23

demand, 81, 242, 244
generation of, 69–71
global production, 232
markets, 81, 174, 187, 252, 253
primary and secondary
   electricity, 27
static electricity, 39
storage, 11, 26, 39, 40
supply, 251
transportation of, 62
electricity mix, 44
electrolysis, 42
electromagnetic waves, 9, 36, 170
emissions trading, 226
energy
   alternative sources, 27, 28, 64, 167,
      192, 235, 246, 249, 268
   biological energy, 4, 12–14, 15, 39,
      45, 47, 126
   chemical energy, 7, 8, 10, 11, 14,
      39, 51, 69, 141, 179
   clean energy, 169, 269
   commercial consumption, 60
   commodities, 26
   concentration of, 11
   concept of, 3, 4, 5, 6
   consumption of energy, 10, 19, 20,
      21, 23, 46, 47, 49, 51, 56, 58,
      59, 60, 62, 72, 101, 103, 104,
      125, 155, 157, 174, 184, 189,
      228, 231, 235, 251, 260, 261
   conversion of, 5, 6–8, 9, 13, 23,
      24, 25, 26, 28, 29, 41, 52, 70,
      71, 74, 118, 128, 166, 170,
      192, 262
   daily requirement of humans
      (table), 49t. 3.2.
   definition of energy, 3, 4
   demand, 103, 118, 141, 146, 148,
      186, 245, 255
   density, 29, 32, 34, 39, 43, 127,
      129
   economics of, 41, 59, 62, 67, 74,
      173, 212, 220–227, 230, 237,
      244, 247, 249
   efficiency, 41, 128, 175, 228, 244,
      260–262, 264
   electrical energy, 39
   energy content of food and fuels
      (table), 48t. 3.1.
   energy loss, 40, 43, 80
   energy release, 71
   'energy transition' (in
      Germany), 117
   environmental impacts of, 175, 223
   external costs of, 117, 222–223
   food energy, 17, 19, 38, 46
   for transportation, 22, 23, 32, 34,
      46, 62, 69, 215, 262–265
   forms of energy, 8–10, 26, 104
   global production, 145
   gravitational energy, 9
   history of energy use, 14–25, 190
   impacts of energy production,
      189–220
   independence, 25, 117, 174, 175
   industrial consumption, 60, 61, 62
   infrastructure, 247, 260, 270, 271
   kinetic energy, 5, 7, 9, 10, 20, 77,
      89, 106, 161
   levels of energy, 10
   markets, 26, 76, 164, 237, 238, 252
   measurement of, 4, 47, 71
   mechanical energy, 8, 10, 19, 70
   nuclear energy. See nuclear power
   politics of, 118, 172, 174, 177,
      236, 261
   potential energy, 10, 40, 77
   primary sources, 41, 44, 82, 93,
      101, 119, 145, 157, 222, 232,
      233, 237, 250, 251
   quantifying energy, 5
   requirement of various human
      activities (table), 49t. 3.3.
   reserves, 63–68
   scenarios for future use, 236, 241,
      246, 253–256, 261
   security, 67, 175
   solar energy, 12, 14, 17, 94, 104,
      118, 155, 160, 178, 235
   sources of energy, 26, 33, 43, 45,
      62, 67, 68, 104, 221, 227,
      228, 230, 233, 237, 246, 248,
      257, 269
   storage, 38–44, 80, 140, 244
   supply, 174, 175
   thermal energy, 7, 9, 11, 20, 41
   transportation of, 12
   useable energy, 11
   wastage of, 141
energy crops, 17, 138, 139, 193,
   214, 216
energy mix, 23, 34, 108, 126, 173,
   193, 233, 238, 246
energy return on investment (EROI),
   167, 228, 229, 230, 248
entropy, 7, 45
Environmental Protection Agency
   (US), 182, 200
environmentalism, 27, 182, 183
European Union, 79, 215, 226
evolution, 13, 14, 15, 17, 74, 177, 181
Exxon Mobil, 143
Exxon Valdez oil spill, 198

Faraday, Michael, 24
Feed-in Tariff Law (Germany), 246
fermentation, 13, 132, 140, 193

Feynman, Richard, 6
fire
  control of, 15, 16
  early uses of, 16
Fire
  control of, 18
first law of thermodynamics, 5, 7, 95
First World War, 24, 31
Fleischmann, Martin, 169
flexible fuel vehicles (FFVs), 139
flood control, 81, 82
Food and Agriculture Organisation (FAO), 141, 239
force, 3, 4, 5, 6, 9
forestry, 141, 145, 146, 148
fracking. *See* hydraulic fracturing
France, 11, 19, 166, 171, 186, 187, 247, 252
Franklin, Benjamin, 39
Franklin, John, 177
Friedman, Milton, 222
Fritts, Charles, 105
Frost, Robert, 6
fuel cells, 27, 42
fuels, 11, 18, 20, 21, 22, 23, 24, 25, 26, 27, 28, 29, 32, 34, 43, 45, 51, 52, 59, 62, 63, 69, 74, 75, 81, 136, 143, 145, 192, 194, 212, 220, 237, 238, 264, 269
  automotive, 12
  for transportation, 139, 146, 173
  fossil fuels, 25, 28, 29, 32, 38, 39, 43, 64, 67, 103, 104, 117, 125, 126, 139, 144, 172, 179, 190, 212, 213, 216, 219, 220, 222, 223, 227, 230, 233, 235, 236, 239, 245, 248, 249, 253
    EROI of, 230
  nuclear fuels, 36–38, 67, 229
  primary and secondary fuels, 27
  solid, 137, 194
  transportation, 143
  wood, 18
Fukushima, 187, 188, 202, 236

gasification
  of biomass, 138
gasoline. *See* petrol
geothermal energy, 68, 73, 148–159, 162, 192, 216, 217, 221, 235, 243, 245, 254, 255, 261, 271
  binary cycle plants, 153, 156
  economics of, 221
  enhanced geothermal system (EGS), 152, 157, 159, 217, 255
  flash steam plants, 153
  ground-source heat pumps (GHP), 155, 157
  impacts of, 216, 217

Germany, 11, 19, 91, 93, 102, 116, 117, 118, 125, 136, 174, 175, 246, 247, 249, 250, 267
Ghawar oil field, 64
Glaser, Peter, 170
global warming, 25, 135, 143, 179, 181, 184, 198, 216, 270
Grand Coulee Dam, 77
Great Pyramid of Giza, 53t. 3.6., 91
Greece, 102
Greece (ancient), 9, 19, 21, 24, 33, 94, 103
greenhouse effect, 178, 178f. 5.1., 179
greenhouse gases, 179, 180, 181, 184, 216, 223, 240
Greenland, 85
Greenpeace, 68, 141, 185, 187, 220, 236
grid parity, 116, 255, 263, 269
Gulf of Mexico, 84, 228
Gulf War, 233

Hall, Charles, 228, 230
heating, 26, 155
  district heating, 152, 162
Hedegaard, Connie, 215
heterotrophs, 12, 14
Hindenburg disaster, 43
Hitler, Adolf, 174
Holland. *See* Netherlands
Hoover Dam, 77, 82
Hopkins, Rob, 268
horizontal-axis wind turbine (HAWT), 88, 90
horsepower, 5, 20
Hubbert Peak Curve, 63
Hubbert, Marion King, 63
Human Development Index, 259
Hume, David, 3
hurricanes, 76, 84, 85
hydraulic fracturing, 35, 152, 201, 235, 247
hydrocarbons, 23, 29, 30, 175, 200
hydrogen, 18, 24, 25, 29, 42, 43, 62, 92, 129, 131, 138, 166, 167, 169, 244, 264
hydrogen bomb, 166
hydropower, 26, 40, 71, 73, 75, 76–82, 108, 116, 165, 177, 185, 192, 206, 228, 233, 235, 243, 244, 247, 256
  early electrification, 77
  economics of, 81
  environmental impact of, 206
  EROI of, 230
  potential of, 82
  pumped storage, 40, 80–81
  run-of-the-river (ROR) plants, 77
  small-scale plants, 77, 79, 80, 271

watermills, 19
waterwheels, 76

Iceland, 56, 149, 217
India, 60, 64, 69, 83, 116, 124, 138, 145, 165, 173, 176, 184, 189, 205, 212, 254
Indonesia, 84, 146, 157, 189, 212, 213
Industrial Revolution, 4, 19, 21, 23, 25, 138, 176, 189, 246
Intergovernmental Panel on Climate Change (IPCC), 180, 181, 241
internal combustion engine, 7, 21, 22, 31, 42, 63, 134, 160, 262
International Atomic Energy Agency, 186, 203
International Energy Agency (IEA), 172, 236, 241, 246, 260
International Thermonuclear Experimental Reactor (ITER), 167
Iran, 117, 174
Iraq War, 233
Ireland, 85, 175, 260
irrigation, 19, 76, 81, 82, 146, 181, 190, 205
Israel, 102, 171, 172
Italy, 19, 117, 118, 119, 120, 124, 141, 157, 187, 207, 249, 250, 251, 252, 256

Jacobson, Mark, 242
Japan, 11, 59, 60, 102, 118, 125, 149, 176, 187, 188, 236, 242, 261
Jordan, 230
Joule, James, 5

Kennedy, John F., 174
Kepler, Johannes, 119
kerosene, 21, 22, 32, 59, 79, 145, 174, 246
Klitgaard, Kent, 230
Knies, Gerhard, 242
Korea, 165
Kump, Lee, 181
Kyoto Protocol, 184, 185, 188, 219, 227

Lao Tzu, 76
Larderello (Italy), 150, 155
Leibniz, Gottfried, 3
levelized cost of energy (LCOE), 221
Leyden jars, 39
lightning, 10, 11, 16
lignin, 18, 126, 132, 141
Lin, Boqiang, 177
liquified petroleum gas (LPG), 145
load profiles (electricity), 72

Lomonosov, Mikhail, 29
Lotka, Alfred, 14
Lovelock, James, 185
Łukasiewicz, Ignacy, 21

Maldives, 196
Mandich, Mitch, 142
Marangoni, Alessandro, 252
Melville, Herman, 21
Mendeleev, Dimitri, 36
metallurgy, 27
methane, 23, 29, 33, 35, 38, 43, 128, 131, 135, 138, 143, 179, 180, 227
Mexico, 145, 157, 171, 176, 197, 212
Middle East, 17, 33, 65, 84, 124, 125, 171, 212, 235, 237, 242
Millennium Ecosystem Assessment, 230, 239, 266
mobile phones, 45, 105, 119
Montreal Protocol, 184
Mossadegh, Mohammad, 171
Mouchout, Auguste, 120, 121
Mount Ararat, 140
Mount Parnassus, 33
Mueller, Alexander, 239
municipal solid waste, 141, 144

Nagasaki (atomic bomb attack), 166
NASA (National Aeronautics and Space Administration), 170
National Ignition Facility (NIF), 167
natural gas, 12, 23, 27, 30, 33, 34, 52, 59, 65, 70, 73, 82, 96, 132, 136, 137, 159, 192, 194, 200, 220, 221, 233, 237, 246, 247, 248, 249, 257
  economics of, 186
  liquified natural gas (LNG), 34, 172
  markets, 172
  shale gas, 34, 35, 67, 173, 199, 201, 235, 237
Nepal, 79, 80
Netherlands, 84, 164
New Zealand, 85, 157, 217, 227
Newton, Isaac, 3
Niger Delta, 197
Nigeria, 80
Nixon, Richard, 174, 175
Northwest Passage, 177
nuclear fusion, 25, 42, 165–167
  cold fusion experiments, 169
  in the sun, 24
nuclear meltdown, 11, 188, 202
nuclear power, 25, 40, 67, 69, 72, 81, 173, 186, 187, 192, 203, 227, 233, 235, 236, 252
  economics of, 62, 113, 187, 247
  efficiency of, 248

nuclear power (cont.)
  environmental impact of, 202
  EROI of, 229
  fission, 29
  phase-out of, 175, 187
  research funding for, 247
nuclear power environmental impact of, 204
nuclear waste, 187, 204, 219
nuclear weapons, 38, 186

ocean energy, 159–165, 243
  'reversed electro dialysis' (RED), 164
  economics of, 164
  impacts of, 217, 218, 219
  ocean current turbines, 162
  ocean thermal energy conversion (OTEC), 160, 162, 165, 217
  osmotic power, 164, 165
  potential of, 165
  salinity gradients, 160, 162
  tidal barrages, 160, 161
  wave power, 160
oil. See petroleum
oil (animal), 21
oil (mineral), 21, 22, 23, 26, 27, 29, 30, 31, 32, 33, 34, 35, 36, 56, 62, 63, 64, 67, 70, 75, 103, 105, 113, 117, 121, 127, 138, 139, 140, 143, 145, 146, 152, 159, 166, 171, 172, 173, 174, 192, 193, 194, 195, 196, 197, 198, 199, 200, 201, 211, 216, 219, 220, 221, 228, 230, 232, 233, 235, 237, 244, 245, 246, 247, 248, 262, 268
  consumption of, 172
  dependence on, 64
  derivative products, 22
  economics of, 64
  exploration, 34, 198
  extraction technologies, 64, 199
  fields, 22
  markets, 172, 236
  nationalization of production, 171
  peak oil, 63, 64
  pricing, 171, 172, 173
  production, 64, 172
  reserves, 63, 64
  spills, 198
  supply, 172, 174
  synthetic, 121
  tar sands, 35, 36, 192, 199, 200, 228, 235, 237
  transportation of, 198
  unconventional, 35
oil (vegetable), 138, 140, 145, 213

oil companies, 171, 172, 211, 219, 227
oil crisis, 25, 27, 32, 120, 139, 170, 187, 233
oil fields, 219
Oil Pollution Act (USA), 197
OPEC (Organization of Petroleum Exporting Countries), 171, 172
  embargo, 172
Oppenheimer, J. Robert, 24, 25
Osborn, Henry Fairfield, 183
ozone depletion, 184

Pakistan, 83
Panama Canal, 177
particles, 9, 10, 36, 169, 193, 194
passive houses, 41, 261
passive solar heating, 261
peak load (electricity), 72, 73
peat, 34, 126, 199
periodic table, 36
Persian Gulf, 22, 64, 171
Peru, 212
pesticides, 56, 216
petrol, 8, 27, 29, 38, 62, 220, 221, 222, 230
petroleum. See oil (mineral)
Philippines, 157
photosynthesis, 12, 14, 38, 94, 103, 106, 126, 130, 179, 193
Pickens, T. Boone, 247
Pifre, Abel, 120, 121
pollution, 43, 59, 132, 175, 176, 179, 182, 193, 194, 196, 198, 200, 203, 204, 207, 216, 219, 223, 226, 248, 266, 269
  from tar sands, 199
Pons, Stanley, 169
population growth, 1, 179, 182, 213, 232, 267
power, 3, 4, 6, 8, 12, 19, 24, 60, 71, 76, 80, 82, 92, 124, 150, 157, 185, 186, 212, 230, 232, 238, 247, 250, 252
  requirement of various devices (table), 50t. 3.4.
power grid, 70, 75, 79, 80, 81, 90, 92, 93, 96, 106, 117, 119, 174, 187, 212, 221, 248, 249, 250, 251, 254, 274
  smart grids, 251
power market, 117
power plants, 10, 38, 40, 45, 47, 51, 52, 62, 69–71, 72, 74, 76, 77, 81, 104, 113, 121, 126, 128, 132, 135, 143, 144, 150, 153, 156, 157, 164, 173, 193, 195, 216, 217, 219, 236, 245, 250, 251